IN9009200
2/2/89

Satellite Remote Sensing for Resources Development

Satellite Remote Sensing for Resources Development

edited by

Karl-Heinz Szekielda

Senior Economic Affairs Officer
United Nations Department of Technical Co-operation for Development

Graham & Trotman Limited

United Nations Department of Technical Co-operation for Development

German Foundation for International Development

Satellite Remote Sensing for Resources Development
Szekielda, K-H. Chapter 1, Figure 1-Estimated Growth of data suitable for digital processing permission for reproduction granted by American Society for Photogrammetry.

The opinions expressed herein are those of the authors and do not necessarily reflect the views of the United Nations.

First published in 1986 by

Graham & Trotman Ltd
Sterling House
66 Wilton Road
London SW1V 1DE
UK

Graham & Trotman Inc.
13 Park Avenue
Gaithersburg
MD 20877
USA

British Library Cataloguing in Publication Data

Satellite remote sensing for resources development.
1. Natural resources—Developing countries—Remote sensing
I. United Nations. *Department of Technical Co-operation for Development*
II. German Foundation for International Development
333.7'028 HC59.7

ISBN 0-86010-805-8
LCCCN 85-82318

Typeset in Great Britain by Bookworm Typesetting, Salford, UK

Printed in Great Britain by
Antony Rowe Ltd, Chippenham, Wiltshire

Contents

Foreword

An Interregional Expert Meeting on the Use of Satellite Imaging RADAR and Thematic Mapping in Natural Resources Development, organized by the Economic and Social Development Center of the German Foundation for International Development — DSE — in co-operation with the United Nations Department of Technical Co-operation for Development — DTCD — was held in Berlin (West) from 21 November to 4 December 1984. As a result of this meeting, the participants made the following recommendations:

A. REMOTE SENSING SYSTEMS AND AVAILABILITY OF DATA

1. Acquisition Platforms and their Continuity

The participants expressed concern over the insecurity which clouds the future of orbital remote sensing platforms — the U.S. Landsat series should be continued, if at all possible. The planned initial ten-year operational lifetime of SPOT is encouraging and received support. ESA/ERS 1, Japan/ J-ERS 1 and Canada's RADARSAT programmes should be given full implementation commitment, as soon as possible, and plans should be developed for system continuity.

The participants noted that development of national and regional remote sensing programmes in developing nations, and establishment and upgrading of appropriate ground receiving stations for these systems depends critically on the prospect of platform continuity.

2. Future Developments

(a) Future developments in microwave remote sensing from space should be encouraged so as to circumvent, among others, the problem of cloud cover and to facilitate extension of application areas. It was emphasized that data from such platforms should be available, on a temporal basis, to all interested nations.

(b) Particular mention was made of the desirability of conducting the repeat flight of SIR-B mission. Broad-based participation, especially including the scientists of developing nations, should be encouraged by NASA. The data gathered over the participating countries should be available to such countries without regard to "investigators' priority", with the understanding that such privileged use of data would be confined to the countries concerned, unless released for wider publication by the SIR-B investigators.

(c) Future development in optical remote sensing from space will produce data products of interest to developing countries such as stereo and advanced array technology. In this context it would be of particular interest to participate in future experimental flights of the metric camera and MOMS sensor family. In order to better satisfy the needs of developing countries, it would be desirable that the United Nations assess the requirements for such participation.

3. Need for Compatibility

The operators of future space systems should ensure compatibility, as much as possible, with operating systems both in terms of ground station and data processing equipment, so as to encourage a larger number of nations to participate in remote sensing programmes and ensure optimum return on their existing and planned investments.

4. Involvement of Developing Countries in Future Systems

(a) The participants urged remote sensing system owners to involve developing countries early enough in the planning and conceptual design stages of their future remote sensing systems to ensure broader and more effective participation.

Such participation would address the special requirements and applications of the developing countries.

(b) The 'Announcement of Opportunity' issued by the space system operators should, in very clear terms, spell out the capability of the equipment involved, and the overall mission objectives. Further, there should be sufficient notice given to compensate for the usual delays encountered in responding to such offers of participation. In this context, special mention was made of the SIR-B Announcement of Opportunity as an excellent example of a comprehensive document.

5. Pricing of Data Products

The real time meteorological data should continue to be available, in the form of direct broadcast services, on a no-charge basis. As for data from land remote sensing satellites, programmes will be operated on cost-recovery basis. The user community should be prepared for changes in the prices of data products. The space system operators should, however, maintain a reasonable level of price structure, and specify such structure to the users as early as possible to allow for necessary budgetary provisions.

6. Cost Sharing

The regional ground receiving stations as well as data bank centres should be financially supported by all of the nations who are serviced by such stations/centres for cost-sharing and conservation of resources.

7. Need for Prompt Supply of Data

It was stressed that ground receiving station operators and data centres should minimize delays in supplying data to users through expeditious data processing and streamlining of procedures. In the majority of cases, the delays are prohibitively long and unacceptable for a large number of application areas. Ground receiving stations and data banks should be established in regions where no such facilities exist at present.

B. TRAINING AND SERVICES

1. Research

(a) Intensive research may have to be conducted in support of remote sensing applications, depending on the needs of a country. As research could be quite capital intensive and time consuming, there is a strong need to share information, to conduct joint research and to avoid duplication, especially in countries whose problems are of a similar nature.

(b) Within this framework, it is important to make the results of research work widely available by means of publications, reports or symposia. It would be helpful to identify international scientific periodicals who would be interested in publishing the relevant results at low charges.

2. Focus on Applications

Special training should be organized for focussing on applications of remote sensing techniques in the priority areas as well as for identifying and developing new applications. Such training could be organized in a country where expertise for the desired application would exist.

3. Optimal Use of Existing Systems

(a) Existing remote sensing resources should be optimized through the employment of the latest techniques such as multi-staging, automated data processing, geographic information systems, multisensor data handling, etc. Training of personnel wherever necessary for achieving this goal should be organized.

(b) To promote training opportunities in an optimum manner, it was considered advisable to designate national remote sensing agencies/space organizations to co-ordinate training programmes in these countries.

4. Technology Transfer

Special technology transfer cells should be established within remote sensing centres. The function of these cells should be to facilitate technology transfer through selection of appropriate technologies, training of personnel, hiring of consultants, planning, phasing and bringing about other necessary interactions.

5. Availability of Specialists in Hardware and Software

There should be regional as well as interregional exchange of specialists both in the fields of the hardware such as ground receiving equipment, data processing equipment etc.; and software such as that for image processing, etc. Training of scientists from the developing nations in these fields should be accelerated to encourage self-reliance.

6. Exchange of Experiences

Developing countries should be encouraged to exchange experiences with regard to scientific management, hardware, software and applications in order to profit from each other's experience and to avoid duplication of effort. To achieve this level of knowledge, these nations should also effectively interact with the developed countries.

7. Formation of Regional Task Groups

Regional task groups should be formed to implement specific activities such as hardware selection, software development, training of manpower, and maintenance of equipment to respond to the needs arising in that region. Each country in the region may specialize in one or more of these activity areas and should notify the regional co-ordinating agency to this effect. To support task groups for the maintenance of equipment, establishment of regional spare parts banks should be considered.

Resolution

The Interregional Expert Meeting on the Use of Satellite Imaging RADAR and Thematic Mapping Data in Natural Resources Development,

• noting the extreme usefulness of the Meeting generally and its obvious impact in achieving the global goal of transfer of technology,

• expresses its sincere appreciation to the United Nations Department of Technical Co-operation for Development (DTCD) and the German Foundation for International Development (DSE) for organizing and playing hosts to the Meeting and resolves to call upon the UN/DTCD to:
— ensure that similar meetings are held at least once a year;
— continue to co-ordinate all activities to bring about these meetings;
— act as a clearing house for all scientific papers from the developing countries to ensure their widest dissemination;
— sponsor those activities as outlined under paragraphs A.1 and A.2 of the Recommendations of the Meeting with relevant agencies and report progress achieved in each case to the Member States.

The DSE will continue to:
— assist the developing countries in the transfer of technology;
— assist in the placement of scientists of developing countries in the universities and institutions of the Federal Republic of Germany;
— provide advanced training opportunities to the scientists of developing countries;
— offer financial assistance for the organization of advanced training, seminars, meetings and purchase of hardware/software within available resources;
— assist in bringing about bilateral arrangements of technical co-operation between the Federal Republic of Germany and the developing countries involving exchange of scientists, providing consultants, and transfer of equipment to execute projects relevant to strengthening science and development in the developing countries.

1

General Aspects on the Use of Satellite Remote Sensing for Resources Exploration in Developing Countries

*Karl-Heinz Szekielda**

ABSTRACT

The paper considers some aspects of remote sensing in connection with resources development in developing countries. Future trends in the technology as well as actual use of remote sensing in resources disciplines and as financial aspects are briefly discussed.

1. INTRODUCTION

Remote sensing has in recent years become an important tool in resource exploration. Although conventional prospecting techniques are still necessary in resource exploration, remote sensing has become an important discipline whereby data over wide areas can be collected and processed within a short time. Remote sensing as applied to resources monitoring includes the basics of spectral data, methods of interpretation and the use or application of these data.

* Address of the Author: Dr Karl-Heinz Szekielda, United Nations Department of Technical Co-operation, New York, N.Y. 10017, U.S.A. The views presented in this paper are those of the author and do not necessarily reflect those of the United Nations.

Compared to mosaics of photographs taken at somewhat different times, the synoptic view from satellite altitudes has the advantage of almost constant illumination over the whole area of investigation. The coverage of large areas from space has certain limitations but on the other hand it is more economical because covering the same area from aircraft altitudes would require, for the same time span, a giant air fleet to substitute the functions of one spacecraft.

So far the application of satellite data in operational development programmes is limited to data obtained in the visible and infrared part of the electromagnetic spectrum. However, with the fast development of sensors, the use of microwave data obtained with spacecraft may be foreseen for the near future. With respect to ground resolution, remote sensing from aircraft is still the most versatile method for natural resources development (Table 1.1), but the analysis of remote sensing data from satellites can be useful in establishing preliminary inventories of resources at a low cost.

Table 1.1. Remote sensing techniques: current measurement capabilities.[a]

		Aircraft	Spacecraft
Electromagnetic	Visible/ Near I.R.	0.03 µm bandwidth 5 m ground resolution	0.1 µm bandwidth 56 m ground resolution
	Thermal I.R.	0.5 µm bandwidth 10 m ground resolution	2 µm bandwidth 600 m ground resolution
	Microwave	multifrequency X,C,L, band multipolarization HH, HV, VH, VV 5 m ground resolution	single frequency L-band single polarization HH 25 m ground resolution
Potential field	Magnetic field	0.1 gamma 10 m ground resolution	6 gamma anomalies over 300 km distance
	Gravity field	none	variable 10 milligral anomalies over 100 km distance (oceans)

[a]NASA-HQER80-2591 (1) 4-30-80.

A great variety of remote sensors have been flown in earth orbit since the first meteorological satellites were launched in 1958: those of potential value to natural resources studies are listed in Table 1.2. The spatial resolution of most of them is, however, inadequate for detailed resources monitoring.

Table 1.2 Sensors flown on satellites with potential application in resource exploration

Satellite	Sensor	Wavelength or frequency	Spatial Resolution
Nimbus I–VII	high-resolution infrared radiometer (HRIR)	thermal i.r.	8 km
NOAA 1–4	scanning radiometer (SR)	visible and thermal i.r.	7 km
NOAA 1-6	very-high-resolution radiometer (VHRR)	visible and thermal i.r.	1 km
Tiros-N	advanced very-high-resolution radiometer (AVHRR)	visible and thermal i.r.	1 km
Landsat 1-2	multispectral scanner (MSS)	four channels, visible and reflected i.r.	70 m
Landsat 3	multispectral scanner (MSS)	visible and thermal i.r.	70 m, 100 m (i.r.)
Nimbus-G	coastal zone colour scanner (CZCS)	six channels, visible reflected and thermal i.r.	500 m
Nimbus 5	electronically scanned microwave radiometer (ESMR)	19 GHz	15 km
Seasat-A	scanning multichannel microwave radiometer (SMMR)	five channels: 6.6, 10, 18, 21 and 35 GHz	15–140 km
Skylab, Geos 3, Seasat-A	short pulse altimeter (Alt)	13.9 GHz	2 km
Skylab, Seasat-A	radar wind scatterometer (Scatt)	13.4 GHz	25 km
Seasat-A	synthetic aperture radar (SAR)	1.4 GHz	25 m
Soyuz-22	multispectral camera (MKF 6)	0.45, 0.54, 0.60 0.66, 0.72, 0.84	about 10 m

The analysis of multispectral photography and other data from space shows that useful information can be derived for application in almost any development of natural resources. Different spectral bands used in monitoring phenomena from above the earth's surface shows that topographic features and surface coverage in the form of vegetation, soil, rocks, etc., can be recorded.

2. GENERAL REMARKS ON REMOTE SENSING FOR RESOURCES DEVELOPMENT

Over the next few years, a number of sensors are scheduled to be flown by the U.S.A., France, Japan, India and the European Space Agency. The sensors on these satellites will continue to provide digital and photographic data for resources investigations. Spann (1980) pointed out that the growth in the amount of available data suitable for digital processing will be particularly dramatic. As shown on Fig. 1.1, these data are expected to increase by more than three orders of magnitude by 1990, to the equivalent of more than 10^8 Landsat scenes.

The extent to which these data will actually be analysed and interpreted for solving 'real-world' problems remains uncertain. At one extreme, these data will be the predominant data type utilized in virtually all research and operational earth resources investigations.

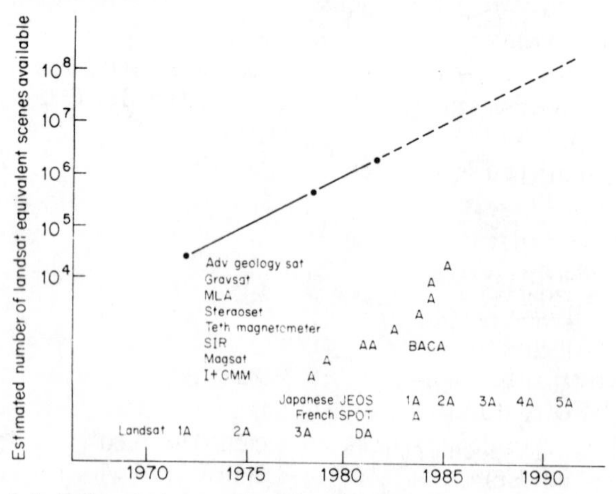

Fig. 1.1. Estimated growth of data suitable for digital processing.

The extent to which foreign countries are utilizing Landsat data was given by Spann 1980 with the following statistics:

- more than 110 countries are participating in some form of remote sensing activity;
- more than 1000 organizations worldwide are involved in remote sensing activities;
- more than 75 countries have utilized Landsat data in various types of resource and mapping studies;
- more than 20 countries and/or international organizations have existing or proposed ground stations capable of receiving data;
- more than 30 countries are classified as having advanced remote sensing programmes; and
- more than 20 United Nations and other international assistance organizations are actively promoting the use of remote sensing data in developing countries or are using this technology in conjuncion with existing development projects.

Landsat images are less detailed than aerial photographs, but possess the following advantages: synoptic scale covering major structures in a single scene; constant raked illumination, providing uniform presentation of features; spectral sensing emphasizing physical contrasts; repetitive imaging, offering seasonal enhancements of features; and global coverage for all land masses between latitudes 81°N and 81°S, which are readily available at low cost. Landsat imagery and aerial photography should be seen as complementary, not competing sources of data and always should be conducted with ground truth programmes.

The acquisition of microwave (radar) data from orbit is now experimentally available, providing images of ground topography in fine detail, an advance of potential value in several areas of natural resource study outside the mineral exploration field.

Landsat imagery is already a routine tool in integrated mineral exploration programmes, and new techniques of interpretation offer improved geological discrimination, which is sure to be of value in the increasingly difficult task of locating buried ores. Available procedures range from simple colour composites of several spectral bands and combinations of Landsat with other data (for example, SLAR and geophysical maps) to facilitate recognition, to complex manipulation of the primary digital data from Landsat computer-compatible tapes. These digital manipulations include contrast enhancement by mathematical transformations such as ratio computation, and classification of rocks or alteration zones by computer matching of characteristic spectral radiances. However, one has to keep in mind that such image analysis techniques are still subject to many geological vagaries if extrapolated over large areas, but showed considerable success for resource applications, particularly in vegetation-free areas (for example, arid and semi-arid regions). Similar techniques may also be applied to discriminate spectral geobotanical anomalies in vegetation-covered areas.

Many new products and services will be needed by users of remote sensing data in the next decade. Some data users require access to satellite data in real-time or near real-time. This trend is evident in several developing countries with the establishment of ground receiving and processing stations. This in turn will increase the demand for communication networks to permit greater use of existing data files and processing analysis models. Also, the demand for independent facilities will increase, a development which can already be seen in the fast development of digital image processing stations using minicomputers.

Capital costs can be minimized by centralizing remote sensing programmes in regional institutions where developing countries and international organizations may use the available facilities. This type of technical co-operation among countries within a region has the advantage of sharing the technology while lessening the financial burden for each participating partner. Remote sensing centres sponsored and developed for regional use are in existence in East and West Africa and others have been established in Asia and Latin America.

The high capital cost of space and ground systems for remote sensing makes it difficult for many countries to adopt their own space programme. For instance, the estimated cost of the Landsat-D and back-up spacecraft programme up to 1977 alone was about $350 million. This included a spacecraft cost of $42.5 million; instrumentation, $85.5 million; ground data handling system $72.5 million; assessment system, $7.9 million; management and administrative support, $36 million; launch vehicle, $18.5 million; and facilities, $3.1 million.

Satellite programmes for remote sensing of earth resources

will certainly expand in the 1980's. The anticipated increases in the use of satellite remote sensing data will result from three evolving trends:

- worldwide data availability through direct readout;
- technological improvements yielding both improved data quality and more types of resource data;
- relative economic austerity.

As pointed out by Spann (1980), natural resources data bases, of which remote sensing data form a part, will become much more prevalent in the 1980's. This will increase the demand for communications/computer networks to provide multiple access to data files and analysis models. This in turn will create a demand for minicomputer- or micro-computer-based stand-alone analyst stations with limited processing capabilities to interface with the larger mainframe computers. However, the most important factors for developing countries in establishing their data bases on natural resources are scientific and technical expertise. Gaps in this expertise have traditionally been filled with training through seminars, fellowship programmes, etc.

The international market is already filled with digital image analysis systems which are specifically designed for the CCT's from the Landsat system. Certainly the data format of planned remote sensing programmes will be different but it is assumed that compatability and transformation of data may be possible in order to use existing facilities or at least to adapt them.

In addition to missions planned in the Soyuz-Salyut and Landsat series, the following missions can be expected for the early to mid-1980's:

- the SPOT system (France, with Belgium and Sweden);
- the marine observation satellites (MOS) and the earth resources satellites (ERS) (Japan);
- the Land Application Satellite System (LASS) and Coastal Ocean Monitoring Satellite System (COMSS) (European Space Agency (ESA);
- the satellite for earth observation (SEO-11) (India).

It appears that the present experimental and the coming pre-operational phases of satellite remote sensing will continue through the 1980's. This will allow, in theory, sufficient preparation to be undertaken by both developing and industrialized countries. However, this preparation, especially by the developing countries, has to be initiated immediately and prior to the launch of the remote sensing system. The past has shown that many years had been lost and the benefits derived from, for instance, Landsat, came much later than one would have expected in 1972.

Table 1.3 The top five countries about which Landsat data were ordered from the EROS Data Center, 1979–1982.

Year	Photographic frames Country	Number	% of grand total
1979			
	U.S.A.	21 890	21
	China	17 580	17
	Australia	5 858	6
	U.S.S.R.	3 601	3
	India	3 549	3
	Subtotal	52 478	50
1980			
	U.S.A.	22 751	21
	China	13 138	12
	U.S.S.R.	6 408	6
	Argentina	4 020	4
	Mexico	3 185	3
	Subtotal	49 502	45
1981			
	U.S.A.	35 506	23
	U.S.S.R.	14 426	9
	China	13 564	9
	Argentina	6 888	4
	Indonesia	6 453	4
	Subtotal	76 837	49
1982			
	U.S.A.	26 132	28
	China	7 445	8
	Canada	5 630	6
	U.S.S.R.	5 340	6
	India	2 443	3
	Subtotal	46 990	51

Country	Computer compatible tapes Number	% of grand total
1979		
U.S.A.	1 304	40
Australia	207	6
South Africa	139	4
Mexico	120	4
Saudi Arabia	87	3
Subtotal	1 857	57
1980		
U.S.A.	1 760	35
China	782	15
Mexico	284	6
Australia	187	4
U.S.S.R.	163	3
Subtotal	3 176	63
1981		
U.S.A.	3 843	39
China	1 014	10
Mexico	530	5
Nigeria	279	3
Indonesia	252	3
Subtotal	5 918	60
1982		
U.S.A.	2 598	52
China	403	8
Mexico	188	4
Australia	158	3
Canada	108	2
Subtotal	3 455	69

Table 1.3 shows that during 1979–1982 only five countries ordered more than 50% of the computer compatible tapes (CCT's) demonstrating that most developing countries were still not in the position to take advantage of a relatively low-cost product for resources development. Similar conclusions can be drawn from sale of CCT's of remote sensing countries and sales of data to countries analysing their own territory (Table 1.4).

There are a number of reasons to expect that the forecast growth acceleration will occur within the next five years. Remote sensing data from satellites will become more readily available than in the past. Other reasons to expect growth are improvements in data quality and an increase in the types of data available. As spatial resolution improves, satellite data more closely approximate the high- and medium-altitude aircraft photographic data that have been used for many years in resources development.

It is not within the scope of this summary to identify the technical specification of future remote sensing sensors. However, the scenario of requirements of users of satellite remote sensing data have to be viewed against the need for data for specific resources development. A study undertaken by Metrics in 1980 showed that a group of non-agricultural resources specialists have an interest in 30–40 metre data, and, on demand, 10-metre data. Furthermore, monthly to seasonal data are all that is required to provide a sufficient frequency of coverage. On the other hand, the group interested in non-renewable resources are far more interested in additional spectral bands in the thermal infrared, with different equatorial crossing times, and in the radar portion of the electromagnetic spectrum.

By far the biggest and up to now unmet need in non-agricultural resources is for worldwide, stereoscopic data. Other groups also indicated a need for stereo, but the geologic community, in particular, considers this their top priority for both exploration purposes and for engineering geologic investigations, including mapping.

Table 1.4 Data sales of CCT's of remotely sensed countries (source: EROS Data Center).

	1975	1976	1977	1978	1979	1980	1981	1982
North America	679	1146	969	1427	1367 2422[a]	1815 3159[a]	3887 7144[a]	2706 3868[a]
Western Europe	54	167	92	153	214 314[a]	228 537[a]	331 832[a]	54 482[a]
Socialist countries of Eastern Europe	13	8	24	27	20 1[a]	199 —[a]	233 16[a]	77 2[a]
Africa	80	212	185	480	456 5[a]	625 4[a]	1671 24[a]	684 51[a]
Middle East	65	133	239	263	188 4[a]	281 26[a]	746 124[a]	168 25[a]
South and East Asia	114	350	141	227	345 43[a]	1081 609[a]	1641 332[a]	654 284[a]
Latin American	77	155	183	171	216 63[a]	531 356[a]	966 326[a]	467 117[a]

[a]By country groups.

3. SPECIAL CONSIDERATIONS FOR COASTAL AREAS

Oceans are major reservoirs of heat and their dynamics and therefore play a significant role in determining global as well as regional climate. Over half of the solar radiation, along with the surface wind stress, is the ultimate energy source for a variety of physical processes in the ocean. The absorption of solar radiation is primarily responsible for the existence of a warm surface mixed layer of the order of 100 metres deep found in most of the world's oceans. The exchange of the ocean's heat with the atmosphere occurs over a wide range of time scales, and largely determines the relative importance of other physical processes in the ocean for climatic change. Some of this heat is used for surface evaporation and is eventually deposited in the atmosphere as latent heat during cloud formation; some is stored in the surface layers; and, some is moved downward into deeper water by various dynamic and thermodynamic processes. The most energetic motion scale in the oceans is that of the mesoscale eddy, whose period is of the order of a few months and whose horizontal wavelength is of the order of several hundred kilometres. With respect to marine resources development most developing countries are focussing on the coastal zone, i.e. open ocean processes will be of minor importance to them at the present time.

The total length of coast in the world is approximately 280 000 miles. Assuming an average width of 50 miles for the coastal areas, of which 50% of this width is land, then the land component of the coastal areas is 12% of the world land area or about 7 million square miles; the waters of the coastal area can be approximated by the shallow seas and waters of the world's continental shelves, which cover an area of 29 million square miles.

This shows that the rapid growth of large-scale activity in coastal areas is taking place in a relatively narrow and limited band. In many industrialized countries, significant proportions of their coastlines have already been developed. For example, 40% of the United States coastline is now exploited, while 75% of the coast of the Netherlands is currently being exploited.

Future important activities will be carried out in the coastal region, especially in those of developing countries:

- large-scale urbanization will continue along coastal belts, with attendant concentration of trade and industry around the major port cities;
- extraction of marine resources, both living and non-living, will concentrate and intensify in shallow, near-shore coastal waters;
- growing international tourism and increasing domestic demand for recreation will place greater demand on seaside areas relative to inland areas.

The multidisciplinary use of coastal areas necessarily requires a well balanced management. Although the management of resources in general has existed for many years, ocean management is a relatively new concept aiming to control:

- the level of activity involved;
- the type of activity involved;
- the resource variable in terms of how a resource contributes to the state of the entire resource system.

If the management and development of marine resources is to be considered, the simultaneous and compatible development of different resources in the coastal region should also be taken into account as to effectively supply, support or provide facilities for the exploitation of resources.

Specific missions for marine-related investigations and resources inventories have been flown only on Nimbus-7, Seasat and the GEOS satellite. Other satellites have also shown the application of infrared, microwave and visible data in the marine field, although they were not built exclusively for this discipline. On an experimental basis, these satellites have been used to collect data for suspended sediment, current boundaries, bathymetrics, ice conditions, surface winds and sea surface temperature (see Tables 1.5 A, B and 1.6). Although the ground resolution of many sensors is still not as good as required, some meteorological satellites with a ground resolution of about 1 km may have an interest to developing countries as can be shown with data form the VHRR and AVHRR. For instance, upwelling areas have been successfully monitored over recent years and have shown that important information on dynamic processes in the coastal region can be gathered at a relatively low cost.

Table 1.5A Marine-related sensors on sun-synchronous environmental satellites.

Name	Sensor	Channels	Derived product	Estimated accuracy
Landsat 3	multispectral scanner (MSS)	0.5–0.6 μm 0.6–0.7 μm 0.7–0.8 μm 0.8–1.1 μm	visible and infrared i.r. signatures of terrestrial, aquatic and nearshore marine regimes	N/A
Landsat D	thematic mapper	0.45–0.52 μm 0.52–0.60 μm 0.63–0.69 μm 0.76–0.90 μm 1.55–1.75 μm 10.4–12.5 μm	same products as landsat-C plus Surface temperature	
ITOS-1 NOAA 2-5	very high resolution radiometer (VHRR)	0.6–0.7 μm 10.5–12.5 μm	day and nighttime cloud cover sea surface temp. (SST)	SST-O.5° C sensitivity to relative changes
TIROS-N Series (NOAA-A THRU G)	advanced VHRR (AVHRR) Note: NOAA Series Channel 10.55-0.68 Channel 5 Only On D, F AND G.	0.55–0.90 μm 0.725–1.10 μm 3.56–3.93 μm 10.5–11.5 μm 11.5–12.5 μm	day and nighttime cloud and surface mapping, surface water delineation SST	SST-0.2° C sensitivity to relative changes

Name	Sensor	Channels	Derived product	Estimated accuracy
NIMBUS-7	scanning multi-channel microwave radiometer (SMMR)	6.63, 10.69, 18.0, 21.0, 37.0 GHz dual polarization	SST sea surface wind speed	≈1.5° C ≈1 m/s
	coastal zone colour scanner (CZCS)	0.43–.45 μm 0.51–.53 μm 0.54–.56 μm 0.66–.68 μm 0.70–.80 μm 10.5–2.5 μm	marine chlorophyll and sediment distribution sea surface temp.	parameters presently under investigation
DMSP Block 5D	operational line-scan system (OLS)	0.41–1.1 μm 8–3 μm	same as Tiros-N	somewhat lower than Tiros-N

Source: J. Sherman, NOAA-NESDIS.

Table 1.5B Marine-related sensors on sun-synchronous oceanographic satellites.

Name	Altitude	Orbit period	Inclincation	Instruments	Data products	Estimated accuracy
Geos-3	843 km	101.7 min 14.2 orbits per day	115°	radar altimeter	marine geoid	± 1–2m
					significant wave height	± 10%
					sea surface topography	± 20 cm
Seasat	790 km	100.8 min ≈14 orbits per day	108°	radar altimeter		
				synthetic aperature radar (SAR)		
				radar scatter-ometer (SASS)		
				scanning multi-frequency radiometer (SMMR)		
				visible i.r. radiometer (VIRR)		

Source: J. Sherman, NOAA-NESDIS.

Table 1.6 Design criteria for seasat.

Measurement			Range	Precision	Resolution or IFOV	Total FOV	Comments
Topography	altimeter	geoid	7 cm–200 m	±10 cm	1.6–12 km SPOT	nadir only	2 grids sampled throughout one year
		currents, surges, etc.	7 cm–10 m	±10 cm	1.6–12 km SPOT	nadir only	along subsatellite track only
Surface winds	microwave radiometer	amplitude	7–50 m/s	±2 m/s, ±10%	121 km SPOT	679 km swath offset 22° from nadir	
	scatter-ometer	amplitude	4–28 m/s	±2 m/s, ±10%	<50 km SPOT	250/500/400 /500/250 km	400 km center block 500 km full winds 250 km high winds only
		direction	0–360°	±20°			
Gravity waves	altimeter	height	1.0–20 m	±0.5 m or ±10%	1.6–12 km SPOT	Nadir Only	along subsatellite track only
	imaging radar	length	50–1000 m	±10%	25 m RESOLUTION	100 × 4000 km	selected coverage (in view of compatible station)
		direction	0-360°	±15%			

Surface temperature	v & i.r. radiometer	absolute	−2° to +35°	±2°C	4 km (IR) IFOV	2127[a] km swath about nadir	(clean air only)
	microwave radiometer	absolute	0° to 35°	±2°C	121 km SPOT	638 km swath offset 22° from nadir	(clouds, light rain)
Sea ice, clouds location and ocean feature	microwave radiometer	Extent	N/A	21 km	21 km	636 km swath offset 22° from nadir	global images, all weather
	v & i.r. radiometer	Extent	N/A	2 km	2 km (visible) 4 km (IR)	2127[a] km swath	broadly sampled images, clean air
	v & i.r. radiometer	Leads	N/A	2 km	2 km (visible) 4 km (IR)	100km	broadly sampled images, clean air
	imaging radar	Icebergs	N/A	±25 m	25 m	100km	real time transmission weather

[a]Data swath width for quoted accuracies is only about 1500 km.
Source: J. Sherman, NOAA-NESDIS.

For several years, the receipt, archive, retrieval and distribution of satellite data was of concern to only a few organizations throughout the world, specifically in the U.S.S.R. and the U.S.A. As a result of the launch and operation of satellites by several other countries and recent increased interest in general in the use of these data, retrospective management of satellite data is being accomplished by many more organisations.

At present there are only two so-called World Data Centres, in Moscow and Washington, D.C., concerned with extra-terrestrial data obtained by rockets and satellites. There is no similar system or forum for earth-oriented satellite RS data managers to increase the awareness on available RS satellite data to the international user community. Data from remote sensing satellites are stored in the following institutions:

3.1 North America

- National Space Science Data Center, NASA, Goddard Space Flight Center, Maryland;
- World Data Center A for Rockets and Satellites, Goddard Space Flight Center, Maryland;
- EROS Data Center, USGS, Sioux Falls, South Dakota;
- National Geophysical and Solar Terrestrial Data Center, NOAA/EDIS, Boulder Colorado;
- National Climatic Center/Satellite Data Services Division, NOAA/EDIS, Washington, D.C.;
- Canadian Center for Remote Sensing, Ottawa, Ontario, Canada.

3.2 Europe

- ESA;
  - ESOC, Darmstadt, Germany
  - Earthnet, Frascati, Italy
  - Telespazio, Rome, Italy
  - Centre de Météorologie Spatiale (CMS), Lannion, France
- Royal Aircraft Establishment (RAE), United Kingdom;
- Dundee University, United Kingdom;
- DFVLR, Oberpfaffenhofen, Federal Republic of Germany.

3.3 South East Asia

- Remote Sensing Technology Centre of Japan (RESTEC), NASDA, Tokyo, Japan;
- Indian National Remote Sensing Agency, Secunderabad, India.

3.4 U.S.S.R.

- Hydrometeorological Center of the U.S.S.R., Moscow/ World Data Center B for Rockets and Satellites.

It has been found that the data stored in these centers could be used more efficiently especially by the developing countries. However, unfortunately, no international mechanisms exist (Landsat data excluded) yet to make countries aware of the quality, information and availability of such stored data.

4. TECHNOLOGY TRANSFER

The most common type of technology transfer promoted by international organizations is carried out through grants or loans to developing countries for the purchase of equipment. The corresponding schematic of this direct transfer is shown in Fig. 1.2A. Advisory services and training are illustrated in Fig 1.2B which depicts interactions between inter-governmental organizations and the less developed countries through information services in the form of data dissemination seminars and training in the use of remote sensing.

The trend in the last few years has been to centralize remote sensing programmes in regional institutions, including the

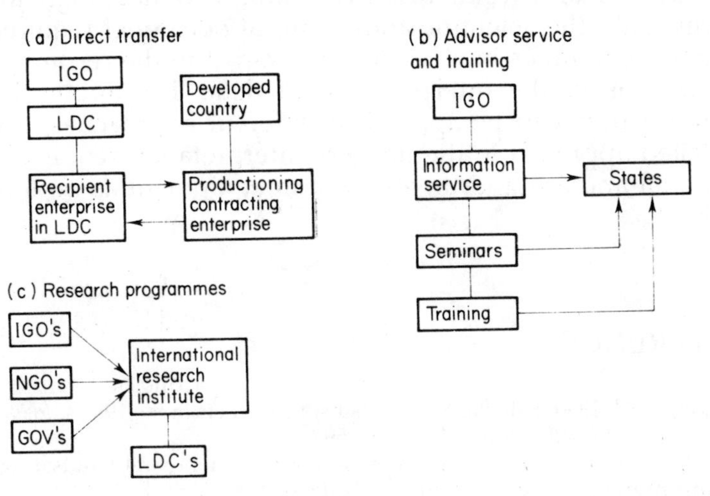

Fig. 1.2. Schematic presentation of technology transfer through international organizations.

interactions between research institutes, the developing countries, inter-governmental organizations, non-governmental organizations and other governments. The generalized function of such research institutes is outlined in Fig. 1.2C. This type of technical co-operation among the countries in a region has the advantage of sharing the technology and also decreasing the financial burden for each participating partner. Remote sensing centres sponsored and developed for regional use are in existence in East and West Africa and others are in preparation for Asia and Latin America.

It is anticipated that the use of satellite remote sensing data will soon enter a very rapid growth phase and will completely revolutionize many of the current data storage, handling, procession and analysis techniques.

With the fast development of new satellite systems and access to data, countries have to face alternatives for the reception and processing of remote sensing data. Most likely, in the future, different space remote sensing systems and receiving stations will be compatible. This in fact gives to most countries, not in possession of space technology, the options to choose their own data source.

Many institutions in industrialized countries as well as in developing countries have the option to start a remote sensing programme through the visual interpretation of satelite data for inventories of natural resources, thus avoiding high investment. Still, the scientific and technical personnel required for such programmes have to be well trained in the resource field as well as in the interpretation of data. The application of Landsat data can be applied in different approaches; a more detailed approach in the imagery interpretation requires more material and a higher investment in equipment as outlined in Table 1.7.

REFERENCES

Spann, G.W. 1980. Satellite Remote Sensing Markets in the 1980's. *Photogrammet. Eng. Remote Sensing*, 46, (1) 65–69

Hilwig, F.W. 1982. Visual interpretation of multi-temporal Landsat data for inventories of natural resources. *ITC Journal* 297–325.

Table 1.7 Instrument facilities and appropriate costs required for Landsat multispectral scanner (MSS) interpretation in relation to the category, purpose and publication scale of the survey.[a]

General category of the map	Purpose of the survey	Publication scale	Landsat materials	Interpretation instruments	Facility and approximate cost
Exemploratory/ schematic	comparison of areas, international level	smaller than 1:1 million	black and white and false-colour composite (FCC) positive transparencies and/or prints	light table; magnifier; mirror stereoscope; slide projector; diazo printer	limited facility budget $US 2 500
Generalized/ schematic	general inventories, national level	1:250 000/ 1:1 million	black and white negative transparencies	in addition to instruments above: colour additive viewer; zoom-transferscope	moderate facility budget $US 25 000
Reconnaissance/ generalized	very large projects, regional level	1:100 000/ 1:250 000	black and white negative transparencies and computer-compatible tapes (CCTs)	in addition to instruments above: facilities for computer assisted	extensive facility budget $US 250 000

Source: J. Sherman, NOAA-NESDIS.
[a] After Hilwig.

2

Present Status of Microwave Remote Sensing from Space with Respect to Natural Resources Monitoring

*Hans Martin Braun and Gerhard Rausch**

ABSTRACT

This paper briefly describes the present status of microwave remote sensing from space. It explains what kinds of radar sensors can presently be used and what their basic operation principles are. The sensors covered are the synthetic aperture radar, the radar altimeter, the microwave scatterometer and the microwave radiometer. The most important space systems (satellites or shuttle missions) carrying these sensors are discussed. Their instrumentation and measurement performance is described. Some examples concerning output products of the described sensor types are given and possible interpretations are sketched. This chapter concludes that the most effective monitoring of resources on earth can be performed by combining data from different sensors (microwave, optic) and that further research and development is necessary within the next decade to establish the most viable remote sensing interpretation procedures.

* Address of the authors: Dr Hans Martin Braun, Gerhard Rausch, Dornier System GmbH, Space Programme Department, P.O. Box 1360, 7990 Friedrichshafen, Federal Republic of Germany.

1. INTRODUCTION

> "If spaceborne synthetic aperture radar (SAR) is to become a viable scientific tool, it must do more than collect interesting pictures."
> R.C. Beal, APL, 1980

This statement of R.C. Beal's describes very clearly the present status of spaceborne radar systems in remote sensing. In other words: the users do not need to have a beautiful toy, they need a highly stable measurement system with accurately known parameters to be able to detect and research unknown target characteristics and to be sure that the measured phenomena are not caused by an unknown behaviour of the sensor itself. This holds for SARs and for all other spaceborne radar systems including scatterometers and altimeters. It shows the importance of calibration in radar remote sensing and it must necessarily lead to an innovative design of calibration methods to guarantee a high degree of confidence in the radar measurements.

For many parts of the world it can be argued that radar

Fig. 2.1. Synthetic aperture radar imagery (7).

systems hold a greater potential than optic systems whether airborne or satellite.

In general, remote sensing has a great potential to enable countries to better understand and exploit their natural resources and to protect their environment.

In countries which are well mapped, where much of thee natural resource base is known and a system of regular statistical returns is available, sensor requirements are directed more towards obtaining detailed elements of data which cannot be observed or obtained by conventional means.

Many countries do not have a similar level of data on their resource base; topographic maps are either non-existant, out of date or unreliable; maps on resources may be small-scale and generalized or localized to areas around existing developments. In the context of world population growth, a greater expectation of life and a general financial recession, countries are facing an urgent requirement to map and develop their existing natural resources.

2. LIST OF ABBREVIATIONS

ALT: radar altimeter (SEASAT)
AMI: active microwave instrumentation (ERS-1)
ATSR/M: along track scanning radiometer and microwave sounder (ERS-1)
ESA: European Space Agency
ERS-1: ESA remote sensing satellite no.1
ITC 2: intertropical coverage zone
JPL: jet propulsion lab
MRSE: microwave remote sensing experiment
NRCS: normalized radar cross section
NASA: National Aeronautics and Space Administration
NASDA: Japanese Space Administration
PDF: probability density function
PRARE: precise range and range rate experiment (ERS-1)
rms: root mean square
RA: radar altimeter (ERS-1)
SEASAT: sea observation satellite
SAR: synthetic aperture radar
SASS: satellite scatterometer (SEASAT)
SMMR: scanning multichannel microwave radiometer (SEASAT)

3. BASIC PRINCIPLES OF MICROWAVE SENSORS

3.1 Synthetic Aperture Radar

A synthetic aperture radar (SAR) produces images of the earth's surface by virtue of the differing local normalized radar cross-section (reflectivity) of the target. As the instrument provides its own source to illuminate the target it can be operated independently of diurnal and weather-influenced solar illumination. Microwave signals in the frequency range of 1–10 GHz can even penetrate heavy rain and hence have an advantage over optical sensors. Spatial resolution is comparable with that of optical sensors and greatly exceeds that of conventional earth-based radars. Radar measurements taken from high altitudes (> 250 km) maintain an almost constant angle of incidence within the swath. As the normalized radar cross-section (NRCS) depends on the angle of incidence, features of signatures from widely separated samples can very easily be compared.

Pulses emitted from the SAR antenna are reflected from a wide swath. All objects within the swath bounce back signals with a strength according to their respective radar cross-sections. The return of a pixel (a resolution element) is characterized by the four quantities of amplitude, phase shift, time of arrival and doppler shift, the last being caused by satellite motion and earth rotation.

The pixel shown in Fig.2.2 is illuminated by many pulses during the SAR overflight and consequently reflects these echoes which are summed up in the same range bin. They differ however by their respective doppler shifts which are known *a priori* by geometry and hence can be calculated.

The doppler shifts are employed by the processor to allocate each partial echo to its proper resolution cell. By inphase addition of all relevant single echoes, a synthetic antenna is created, the beamwidth of which is so narrow that it illuminates only one resolution cell whose echo represents a pixel. In-phase summing of partial echoes requires that all signals generated in the SAR be coherent.

In general all resolution cells can be localized by virtue of a quasi-orthogonal grid iso-range and iso-doppler curves at each moment. The desired spatial resolution on the ground is acquired in one direction (via known geometry) by time of

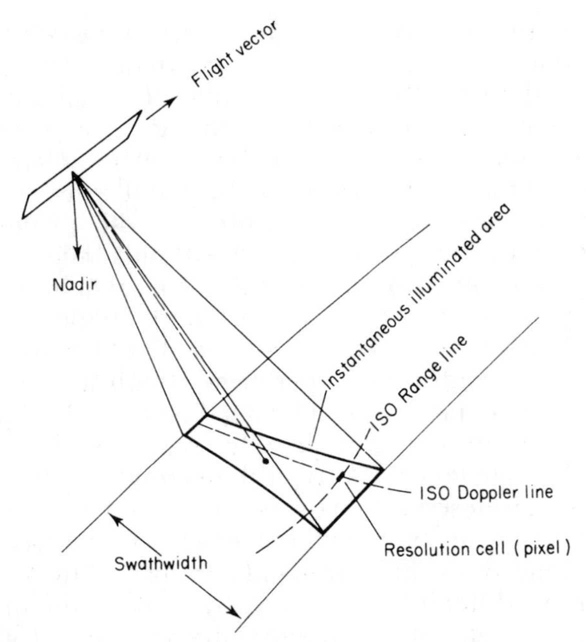

Fig. 2.2. SAR measurement geometry.

arrival of the echo (range gates) which determines the allowable pulse duration. In the other direction an azimuth resolution in the order of one half the antenna length (along flight path) can be obtained. This is determined by the attainable doppler resolution and hence by the time duration of illumination for a point target (integration time). If this fine resolution in azimuth is not desired, several azimuth cells can be summed incoherently after the general correlation in the processor, whereby the picture quality is enhanced in terms of grey level resolution.

If a conventional radar were to yield an azimuth resolution of about 5 m it would have to employ an antenna length of 1.25 km! In other words, the SAR produces a 'synthetic aperture' of dimensions unfeasible with a 'real aperture' by use of a conventional real antenna of less then 20 m.

3.2. Microwave Scatterometer

A microwave scatterometer can ideally be used to measure the wind vectors over the ocean surface.

The measurement principle of a microwave scatterometer derives from the fact that at microwave frequencies, the ocean surface roughness, which is a function of actual wind conditions, appears like a reflection grating. This results in a functional dependence between the normalized radar cross-section ($\delta°$) of the ocean surface and the wind speed.

The radar cross-section is anisotropic with respect to the angle between wind vector and incident radar beam. With the aid of several $\delta°$- measurements of the same area from different measurement directions the actual wind vector in terms of speed and direction can be determined. The conversion of the $\delta°$ value into wind data is performed with a mathematical model, which defines the relationship between $\delta°$, wind speed, wind direction, incidence angle of the scatterometer pulse and the signal polarization. A typical microwave scatterometer configuration is described below (see Fig. 2.3).

The scatterometer shown illuminates the sea surface sequentially for reflectivity measurements by radio frequency (RF) pulses from different directions by three antennae. The nominal look angles of the antennae are 45° fore and aft as well as broadside to the satellite velocity vector. Reflectivity data will be provided for a wide continuous swath along the satellite

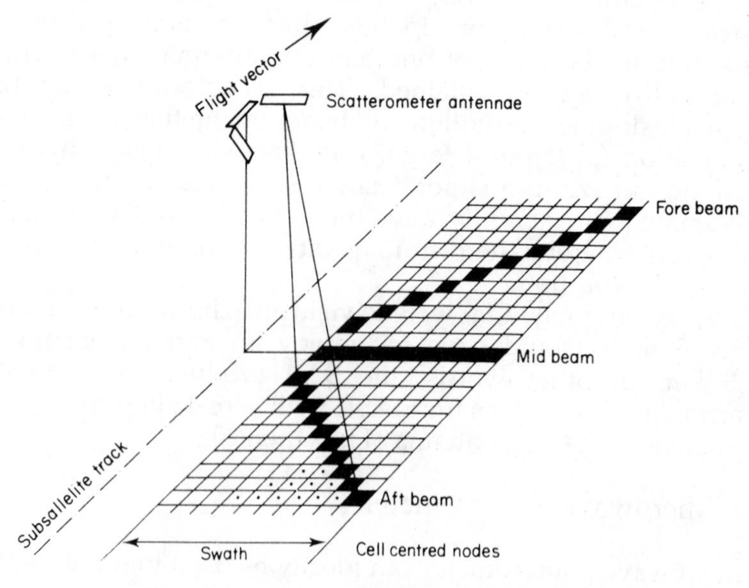

Fig. 2.3. Microwave scatterometer geometry.

ground track to deduce wind speed and direction at notes (resolution cells) along and across the sub-satellite track within the swath.

Each node is centred within a resolution cell that is determined in range direction by appropriate range or doppler gating of the received echo signal and in azimuth by averaging of corresponding echo signals being limited by the azimuth beamwidth of the antennae.

Proper processing extracts the noise from the echoes to provide high radiometric accuracy and determines the ocean surface wind vectors by correlating the three reflectivity values per node with a model function, which has been evaluated and updated now for more than one decade.

3.3 Radar Altimeter

The radar altimeter is a nadir-looking active microwave instrument that can be operated over ocean, ice and land.

Over ocean it is used to determine the significant waveheight (SWH), the wind speed, and the mesoscale topography. Over ice it is used to determine the ice surface topograhy, ice type, and sea/ice boundaries. Both types of operation, sea and ice, have been used for years in several space projects. Operations over land, however, require an extremely high standard of sensor technology and are reserved for advanced future projects. The following example concerns operations over ocean and describes the basic measurement principle.

The ocean radar altimeter is based on the electromagnetic backscattering of the ocean surface in response to a narrow pulse under near normal incidence. In fact the average return echo (average over many return pulses) carries information on the sea surface such as the mean sea level and the sea state (significant wave height and wind speed). The above power response can be modelled by the convolution of the following three time functions:

- the average flat surface impulse response (decreasing exponential with a time constant related to the antenna beamwidth);
- the height (converted into delay time) probability density function (pdf) of the sea surface scatterers (gaussian with standard deviation proportional to the rms waveheight);

- the altimeter system point target response (gaussian with standard deviation proportional to the pulse width).

The resulting function is a shaped pulse, see Fig. 2.4 whose leading edge has a slope proportional to the significant wave-height, and whose power carries information about sea reflectivity, which in turn depends on the wind speed.

Further ocean information — the mean sea level — is conceptually a classical measurement of radar systems, as it is related to the time delay between transmitted pulse and the half power point of the received echo.

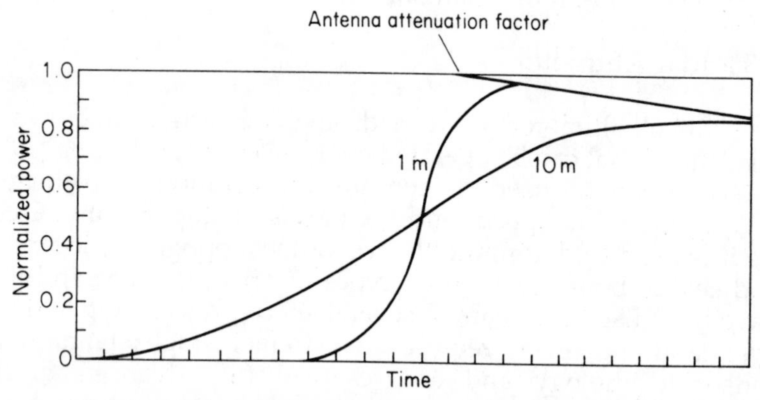

Fig. 2.4. Average return power waveform for two rms waveheights.

3.4. Microwave Radiometer

A microwave radiometer performs passive microwave observations of the atmosphere measuring the thermal emission from various broad layers in the atmosphere, the altitude of which can be determined via radio frequency. Physical processes which typically determine the observed emissions include:

- line emission and absorption by molecular species such as O_2, H_2O and O_3, which typically are in thermal equilibrium at altitudes below ~ 100 km, and for which departures from thermal equilibrium can become interesting at higher altitudes;

- pressure broadening of spectral lines, which typically dominates line shape above ~ 50 km and which requires use of a matrix equation of radiative transfer;
- nonresonant absorption and scattering by droplets small compared to a wavelength and sometimes by water or ice particles which are not so small and which may have nonuniform compositions and non-spherical shapes that scatter radiation in particular directions;
- Doppler thermal broadening, and frequency shifts due to wind; and
- surface effects.

Using scanning antennae the three-dimensional distributions of atmospheric temperature and composition can be estimated by mathematically operating on the microwave spectrum observed. In this process infrared, weather forecasts, or other data are sometimes incorporated to correct the microwave observations.

4. MAJOR REMOTE SENSING SATELLITES/MISSIONS

4.1 SEASAT

NASA's SEASAT (see Fig. 2.5) was the first satellite dedicated to studying the ocean surface. It was launched on 27 June 1978 (European time) into an orbit of about 800 km high and it died by a power failure after little more than 100 days in operation. This satellite brought together a suite of four microwave sensors. They were a radar altimeter (ALT), a wind scatterometer (SASS), a scanning multi-channel microwave radiometer (SMMR), and last but not least a synthetic aperture radar (SAR). This microwave payload was supported by a visible and infrared radiometer.

4.1.1 *Synthetic Aperture Radar*

The main purpose of the SEASAT SAR was to obtain high-resolution imagery of the sea surface and sea ice. The SAR operated at L-band (1.4 GHz) and was designed to image a 100 km swath offset some 250 km to the right of the satellite's sub-orbital track. The geometric resolution was 25 m and the respective SAR data rate was so high that no on-board

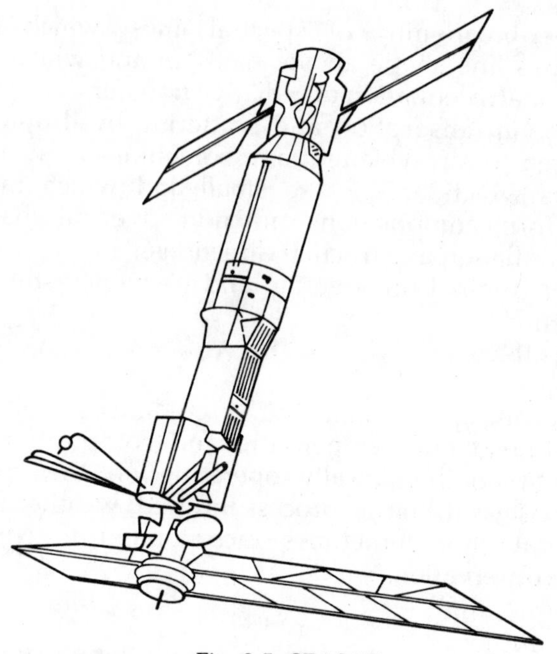

Fig. 2.5. SEASAT.

recording was possible and the raw data were transmitted via a real analog channel to a ground station.

The measurement objectives of the SEASAT-SAR were:

- to obtain radar imagery of ocean wave patterns in deep oceans;
- to obtain ocean wave patterns and water–land interaction data in coastal regions;
- to obtain radar imagery of sea- and fresh-water ice and snow cover.

4.1.2. *Radar Altimeter (ALT)*

The SEASAT ALT was a third generation instrument, previous models having been flown on SKYLAB and GEOS-3. It was designed to measure altitude to 10 cm rms in order to detect currents, tides, storm surges, and to refine the geoid. It operated at a frequency of 13.5 GHz, its data rate was 10 kbps, and it covered a footprint (spatial resolution) of 2–12 km depending on the sea state. It provided an estimate of the significant waveheight of 0.5 m or 10%.

The primary measurement objectives of the SEASAT ALT were:

- to measure very precisely the satellite altitude above the sea surface;
- to measure the significant wave height of the ocean surface at the subsatellite point;
- to extract oceanographic and marine geoid information.

4.1.3. *Wind Scatterometer (SASS)*

SEASAT-SASS was a sensor that had flown previously on SKYLAB.

The scatterometer was designed to measure ocean surface wind speed from 4 m/s to > 26 m/s with an accuracy of 2 m/s or 10%. It operated at Ku-Band (14.6 GHz). Four antenna beams arranged at 45° to the spacecraft's track were deployed to estimate wind direction. SASS was designed to sweep out two 500-km swaths separated by 400 km, together with a narrower swath around the sub-satellite track. Its resolution cell was about 50 × 100 km.

SASS could transmit and receive at both vertical and horizontal polarization. The switching of this facility between the four antennae gave the instrument a total of nine operating modes which were selected according to the nature of the experiment and the quality of the available surface data. On several of those modes coverage was restricted to one side of the space craft.

The algorithms being used to extract the wind vectors from the radar echoes were developed prior to SEASAT's launch and were tested against the *in situ* recordings of calibrated anemometers during the specially-commissioned Gulf of Alaska SEASAT experiment (GOASEX) and, subsequently, against similarly well-calibrated surface measurements in JASIN.

The primary measurement objectives of SASS were:

- to deduce local wind vector information from ocean radar scattering coefficient information;
- to obtain synoptic ocean radar scattering coefficient measurements over a wide variety of sea and weather conditions and instrument parameters;
- to obtain radar scattering-wind vector interaction data over the ocean.

4.1.4. *Scanning Multichannel Microwave Radiometer (SMMR)*

The SMMR, the only passive sensor of SEASAT's suite of four microwave instruments, covered a 600-km swath to the right of the spacecraft. It received both vertically and horizontally polarized radiation at five frequencies: 6.6, 10.7, 18.0, 21.0 and 37.0 GHz. From this matrix of ten channels of information the main goals for the instrument were the extraction of:

- sea surface temperature to 2°K absolute and 1°K relative over resolution cells approximately 80 (multi) × 150 km;
- ocean surface wind speed (but not direction) to 2 m/s;
- integrated liquid water content and water vapour;
- rain rate;
- ice age, concentration and dynamics.

SMMR is the only of SEASAT's now functioning on another satellite, NIMBUS-7, so that the complex algorithms and their subsequent refinement during validation experiments can be applied to an operational instrument.

The SMMR measurements also provide important information being used to correct errors in the radar altimeter data due to atmospheric effects.

4.2. Microwave Remote Sensing Experiment (MRSE)

The development and the qualification of the spaceborne Microwave Remote Sensing Experiment (MRSE) (Fig. 2.6) was the first European step into radar remote sensing from space. Dornier System has designed, developed and qualified this experiment for the first spacelab flight, scheduled for Autumn 1983. The MRSE contained radar sensors of three kinds, a synthetic aperture radar (SAR), a dual frequency scatterometer and a microwave radiometer. This instrument failed on its first flight before SAR measurements could be made; hence only few radiometer and scatterometer data are available. These are presently under evaluation. A re-flight of the repaired MRSE is planned as soon as possible. The measurement frequency of the MRSE lies within the X-band. This makes the SAR mode of the MRSE very interesting, because it will perform the first X-band SAR measurements from space for civil purposes.

Fig. 2.6. Microwave remote sensing experiment (MRSE).

4.3 Shuttle Imaging Radar

In November 1981, the Shuttle Columbia carried the Shuttle Imaging Radar (SIR-A) which successfully acquired radar images over numerous areas of the world. The objective of the experiment was to assess the capability of spaceborne radars for geologic mapping and Earth resources observation.

The SIR-A is a synthetic aperture imaging radar which uses the motion of the platform to synthesize a long aperture which then would allow high resolution imaging. The SIR-A achieved a resolution of 38 metres over an imaging swath of 50 km. The geometry was selected to achieve 50° incidence angle, to acquire data where the scattering mechanisms are mainly due to surface roughness. This allows the comparative analysis between the SIR-A images and the SEASAT SAR images (acquired in 1978).

SIR-A was the first of a series of Shuttle-borne radar missions followed by SIR-B, which was flown a short time ago in October 1984. The next flight, SIR-C, is planned in 1988.

4.4. ESA Remote Sensing Satellite No.1 (ERS-1)

A big step towards operational radar systems in space is the development of the ERS-1 to be launched in 1988. Under leadership of Dornier System a large industrial consortium has designed the first ESA Remote Sensing Satellite (ERS-1) on behalf of the European Space Agency (see Fig. 2.7). This first operational radar satellite in Europe will perform the following geophysical measurements:

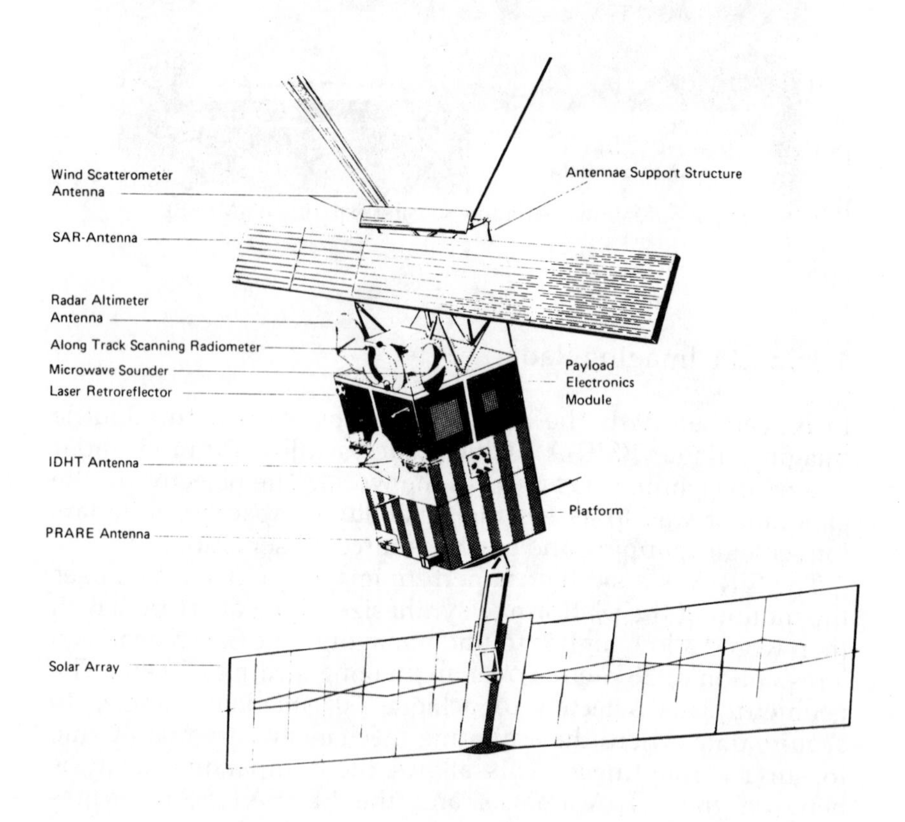

Fig. 2.7. ERS-1

- imaging of oceans and land by a synthetic aperture radar (SAR) having 30 m spatial resolution and more than 80 km swath width;
- ocean wave spectra determination by a Fourier spectrum analysis of 5 × 5 km samples of an ocean SAR imagery being sampled with an interval of 100 km in flight direction;
- wind vector field measurements over the oceans by a microwave scatterometer (C-Band) having the objective of measuring windspeed with a precision of 2 m/s or 10% and wind direction with a precision of 20 degrees within a range of 4-24 m/s windspeed;
- significant waveheight (SWH), windspeed and spacecraft altitude over the oceans shall be determined by an altimetre operating within Ku-band. The measurement precision objectives are: 10 % of SWH, 0.01 m of altitude and 0.5 dB of δ° which is a measure for the windspeed.

It is obvious that these measurement objectives need a sensor design with a high grade of optimization. The SEASAT performance evaluation campaigns taught us that such measurement accuracies are partly feasible, but on the other hand they gave us the impression that it is very difficult to guarantee such high measurement precision under all circumstances. It has been concluded that this problem of precision can only be solved by considering advanced system calibration methods. Some of these methods are based on the use of radar ground-based transponders, receivers or transmitters.

The ERS-1 carries two primary active microwave instruments: a so-called active microwave instrumentation (AMI) and a radar altimeter. (RA).

The *active microwave instrumentation* (AMI) is an active C-band microwave sensor package capable of performing three distinct functions corresponding to different measurements and output products:

- *synthetic aperture radar (SAR)* for high quality wide swath imaging over ocean, coastai zones and land. The spatial resolution is 30 m over a swath of approximately 80 km. The radiometric resolution is 2.5 dB at 30 m and 1 dB at 100 m geometrical resolution. The minimum δ° is -18 dB.

The incidence angle on the ground is 23° at mid-swath. The polarization is VV;

- a *wave mode* using the SAR for determination of two-dimensional ocean wave spectra through Fourier Transform of 5 km × 5 km SAR images. The wave mode will be operated each 100 km along track, and it may be interleaved with the wind scatterometer measurements. The measurement objective is to derive wavelength samples in the range from 100 to 1000 m with at least 12 steps. Angular samples shall be derived from 0 to 180° azimuth in at least 30° steps. The spectral energy density measurement objective of each sample is to have more than 20% accuracy;
- a *wind scatterometer mode* for measuring wind speed in a wide, continous swath up to 500 km over the ocean. The measurement objective in geophysical terms is given by a wind speed range from 4 to 24 m/s with a tolerance of 2 m/s or 10 %, whichever is larger. The wind direction determination tolerance is 20° of true wind direction. Spatial sampling for final wind velocity data is to be better than 50 km with a regular sample grid of 25 km spacing on the ocean surface.

All three AMI measurement modes require absolute and relative calibration with a very high accuracy. The SAR mode calibration requirement is 3 dB absolute and 1 dB relative. The wind mode calibration requirement is 0.35 dB in relation to the insensitivity of the wind model to high windspeeds.

The major AMI technical characteristics resulting from the design within ERS-1 definition phase are given in the following table (Table 2.1) for the three measurement modes.

The *radar altimeter (RA)* is a nadir-looking active microwave instrument which is operated over the ocean and over ice. Over the ocean it is used to determine the significant waveheight, wind speed and mesoscale topography. Over ice it is used to determine the ice surface topography, ice type and sea/ice boundaries.

The microwave measurements comprise the time delay between transmission and reception of a pulse, the slope of the leading edge of the return pulse, the amplitude of the return pulse, and the echo waveform. These measurements are used as follows:

- the altitude is determined from the measured delay time after correction of propagation delays caused by ionosphere and troposphere; these corrections can be taken from measurements made by the Along Track Scanning Radiometer/Microwave Sounder (ATSRM) and the Precise Range and Range Rate Experiment (PRARE). Absolute calibration will be performed by using the laser retroreflector during zenith overflights over a laser ranging station;
- the precision requirement on time delay measurement over the ocean is that 68% of measurements shall lie within ± 0.66 m/s of the fitted mean for sea states up to 20 m significant waveheight (averaged over one second). Over ice a factor of four lower precision is required due to a larger time window;
- the significant ocean waveheight will be calculated from the slope of the leading edge of the return echo. The accuracy

Table 2.1 Major technical characteristics of the ERS-1 active microwave instrumentation.

Modes	Imaging	Wave	Wind
Measurement frequency		5.36 GHz	
Measurement principle	SAR	SAR	scatterometer
Pulse width	31.1 μs	12.3 μs[a]	50-130 μs
Peak transmit power		4.8 kW	
Pulse repetition freq.	1640–1780 Hz	1640–1780 Hz	80–250 Hz
Planar array antennas size	10 × 1 m	10 × 1 m	fwd/aft 3.6 × 0.38 m mid 2.3 × 0.38 m
Waveguide material	metallized CFRP		aluminium
Total mass		356 kg	
DC power consumption	1262 W	236 W	438 W
Data rate (buffered)	100 MBIT/s	0.8 MBIT/s	65 KBIT/s

[a]This is presently under review.

requirement is 0.5 m or 10 %, whichever is greater in a range of sea states from 1 to 20 m SWH;
- the wind speed over sea surfaces will be estimated from the power level of the backscatter signal; furthermore the location of sea/ice boundaries can be derived. The measurement accuracy requirement is ± 0.5 dB in a range of δ° between -2 and +40 dB.

In addition the instrument will provide echo waveform measurements averaged over 50 m/s.

The technical characteristics of the ERS-1 radar altimeter which resulted from the definition phase are given in the following table (Table 2.2).

Table 2.2 Major technical characteristics of the ERS-1 radar altimeter.

Measurement frequency	13.7 GHz (K_U band)
Measurement principle	full deramp concept
Chirp length	20 μ s
Bandwidth	330 MHz
Peak transmit power	50 W
Pulse repetition freq.	1 KHz
Onboard signal processor	suboptimum maximum likelihood estimator
Paraboloid antenna	1.2 m diameter
Total mass	96 kg
DC power consumption	134.25 W
Data rate	15 KBIT/s

4.5 Other Spaceborne Remote Sensing Programmes

In this section some other remote sensing programmes are listed and briefly described:

- AERS (ERS-3): the European Space Agency plans the development of a Land Application Remote Sensing Satellite AERS following the ERS-1 and a mostly identical ERS-2. It will carry an advanced SAR sensor. Its launch is planned after 1991;
- GEOSAT: the U.S.A. Navy plans a late 1984 launch of GEOSAT (Geopotential Satellite). The satellite carries a

Seasat-type radar altimeter to determine the marine geoid — the mean topography of the ocean. The altimetry data will also provide information on mesoscale fluctuations in ocean surface topography, surface wind speed and wave height. GEOSAT will be in an 800 km orbit, and is designed with axial symmetry to minimize orbital variations due to atmospheric drag effects. It is gravitationally stabilized by a large (50 kg or 110lb) mass on the end of 7 m (22.4 ft) boom;

- Japanese ERS-1: NASDA (Japan) Satellite with a Microwave Scanning Radiometer to be launched in 1986;
- NIMBUS 7: the last of NASA's Nimbus series of satellites dedicated to instrument development, NIMBUS-7 was launched on October 23, 1978, and continues to operate in 1984 for its sixth year. The design of the NIMBUS-7 spacecraft was based on the Landsat series. Two of the eight instruments on board are ocean sensors — a SMMR (virtually identical to the SEASAT SMMR) and the Coastal Zone Color Scanner (CZCS);
- NROSS: the U.S.A. Navy plans a 1988 launch of the Navy Remote Ocean Sensing System (NROSS) satellite, which again builds on the SEASAT/NIMBUS-7 heritage. The payload includes a Seasat-type radar altimeter, an improved (six-antenna) scatterometer provided by NASA, a multi-frequency microwave imaging radiometer (SSM/I) similar to the SMMR, and a lower frequency microwave radiometer. Observations of sea surface wind velocity, wave height, temperature, sea ice concentrations and location of oceanic fronts and eddies, will provide tactical Naval support and will be used in the Navy's operational forecast;
- Radarsat: this is a Canadian Satellite for Microwave Remote Sensing primarily equipped with a Synthetic Aperture Radar. The launch is planned for the end of this decade;
- TERS: this is a remote sensing satellite of The Netherlands and Indonesia which will be launched in 1990 (?) and will carry an SAR;
- TOPEX: NASA's TOPEX (Ocean Topography Experiment) satellite is proposed for an early 1989 launch into a 1300 km orbit, with an inclination of about 65°. It will carry a very precise radar altimeter and orbit determination packages so that the surface topography of the ocean can be measured to an accuracy of 14 cm or less across an entire ocean basin.

TOPEX data will provide an opportunity to significantly increase the understanding of large scale oceanic circulation, and augment the planned World Ocean Circulation Experiment.

5. DATA PRODUCTS AND THEIR INTERPRETATION

5.1 Wind Scatterometer

Sea surface winds, the fundamental driving force for ocean waves and currents, also profoundly influence the exchange of heat between the atmosphere and the seas. On a global scale the winds and ocean currents are equal partners in redistributing the excess solar heat gained in the tropics to the cooler polar regions. Since both the oceans and atmosphere transport about half of the heat from the equator to the poles, a knowledge of their dynamics is crucial to understanding our climate.

Our picture of ocean winds has come primarily from information collected by ships. Because ships generally follow well travelled sea routes, our knowledge of marine winds has been confined to limited areas. Vast ocean regions remain unexplored and, consequently, no regular data are available to calculate seasonal variations.

In 1978, the scatterometer (SASS) on the SEASAT satellite obtained the first global measurements of ocean wind data. The image in Fig. 2.8 is the first comprehensive view of the winds over the world's oceans and was constructed from data collected by SEASAT. Seven orbits over the world's ocean during a 12 hour period on September 14 and 15, 1978 were used. Higher windspeeds are shown in white; lower in grey. White streamlines that parallel the wind flow have been interpolated across areas not covered by the satellite path. Over the tropical Atlantic, the 'easterlies' (blowing from east to west) or trade winds coverage north of the equator forming the Intertropical Convergence Zone (ITCZ) indicated as thick black line. Intense storms originating in the ITCZ are called hurricanes when in the Atlantic.

Over the Indian Ocean the monsoon winds, which reverse direction seasonally, blow from the southwest along the coast of Equatorial West Africa and West India at this time (3). In the mid-latitudes (30° to 50°) of both hemispheres, the prevailing

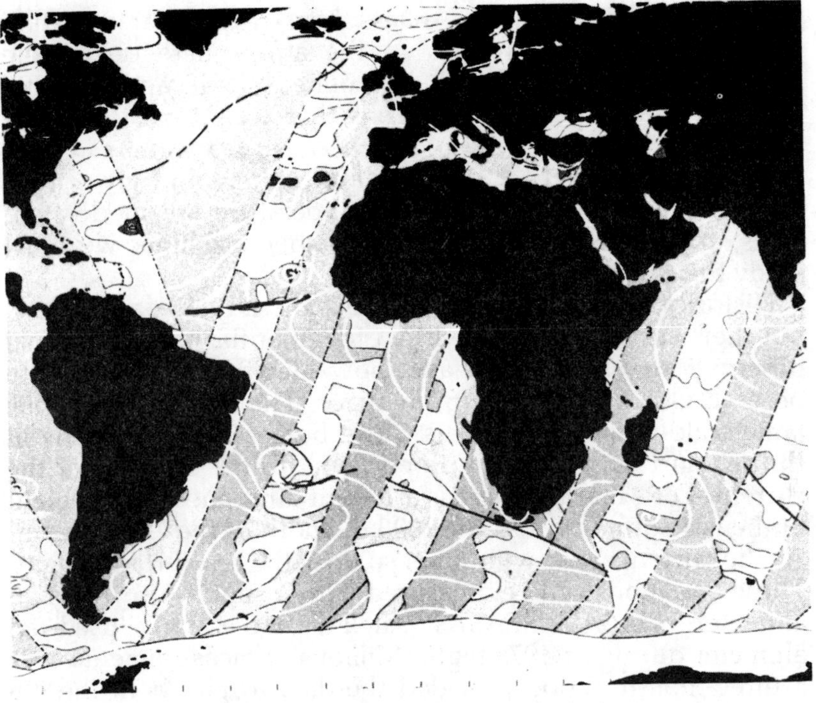

Fig. 2.8. Global marine winds (1).

winds are from the west. These mid-latitude 'westerlies' exhibit much more variability than the tropical easterlies. The westerlies are dominated by rotating wind systems which are associated with areas of high and low pressure. In the Northern Hemisphere, the low pressure systems rotate counter-clockwise and indicate stormy weather; high pressure systems have a clockwise rotation. In the Southern Hemisphere the direction of rotation of high and low pressure systems is reversed. Much of the South Polar Sea is still frozen in September and is shown in white. This ice edge was also determined from the scatterometer data.

5.2. Radar Altimeter (1)

The circulation of the oceans has many direct consequencies for life on Earth. Without the ocean, large areas of our planet would be unbearably hot or cold. The oceans modulate global temperatures in two ways. First, warm ocean currents carry to the poles nearly half the excess heat from solar radiation accumulating in the tropics — the atmosphere carries the remaining half. This redistribution of heat significantly reduces the extreme temperature contrasts that would exist between the equator and poles. Secondly, because sea water can hold considerably more heat than the atmosphere and gives it up slowly, seasonal fluctuations in temperature are moderated. Thus, to understand the global weather machine, we must study the ocean's circulation.

Historically, ships have been used to study ocean currents, but their surveys are limited to a few months and to regional observations. Scientists have now devised techniques to observe global circulation from space. The concept is simple: large-scale ocean movements cause bulges or depressions in the sea surface which are proportional to the strength of the current. For example, there is a 100 cm difference in the height of the sea surface across the 100 km width of the Gulf Stream. As the currents vary with time, so does the sea surface height. These variations can be measured with a sensor known as an altimeter. NASA's SEASAT satellite successfully used an altimeter during its 1978 flight. Millions of measurements over a three-month period provided the data to plot both the sea surface topography and ocean circulation seen in this image (see Fig. 2.9).

The height of the sea surface relative to the centre of the Earth is not only a function of ocean currents, but also of the Earth's composition. Changes in composition have an effect on the Earth's gravity field and are reflected in the relief of the sea

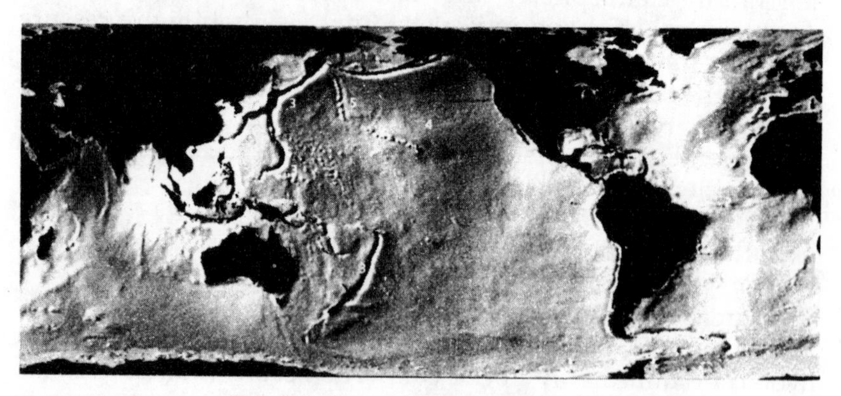

Fig. 2.9. Sea surface topography (1).

surface. This map of the average sea surface topography — the marine geoid — was produced from 70 days of SEASAT Altimeter data. The results clearly show the relationship between the ocean surface and the changes in gravity caused by the underlying ocean-bottom topography. Since the ocean surface predominantly follows the Earth's geoid, this dramatic image is especially useful for charting poorly surveyed areas of the world such as the Southern Ocean surrounding Antarctica. By mapping the ocean surface from space, scientists obtain valuable information about topography and composition of the ocean floor. For example, over a submarine trench, the geoid depresses the surface as much as 60 m closer to the centre of the earth. In contrast, seamounts can cause the surface to bulge as much as 5 m above average sea level.

This image, (Fig. 2.9), which has a spatial resolution of about 50 km was computer generated and the changes are revealed as if the map were illuminated from the northwest. Seen here are the characteristic features of the ocean floor: the mid-ocean ridges, trenches, fracture zones and seamount chains. Clearly visible are the mid-Atlantic ridge (1) and associated fracture zones (2), the trenches along the west and northwest margins of the Pacific (3), the volcanic Hawaiian Island arc (4) and the Emperor seamount chain (5).

5.3. Microwave Radiometer [1]

As the sun migrates annually between hemispheres, the atmosphere, land and ocean system responds with annual temperature variations. While the atmosphere and land experience enormous temperture changes in the high and mid-latitudes, the oceans remain more constant. This is due to the high heat capacity of water relative to that of air and land. Without the oceans, the Earth's temperature would fluctuate radically. Thus, the waters covering 70% of our planet's surface act as a massive thermostat which moderates our global climate. Conversely, small changes in ocean temperature patterns can result in dramatically altered global weather.

Monitoring global temperatures, especially from the oceans, has traditionally been impossible because of the lack of data from many areas. Now, satellite sensors are used to observe month-to-month and year-to-year changes in surface temperatures. Examples are these images (Fig. 2.10) produced using

JANUARY

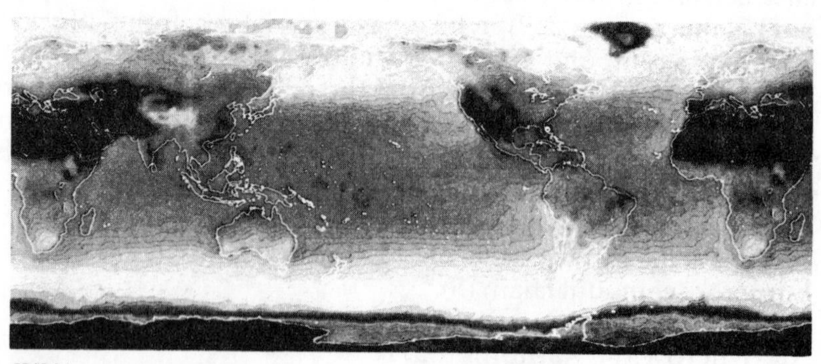

JULY

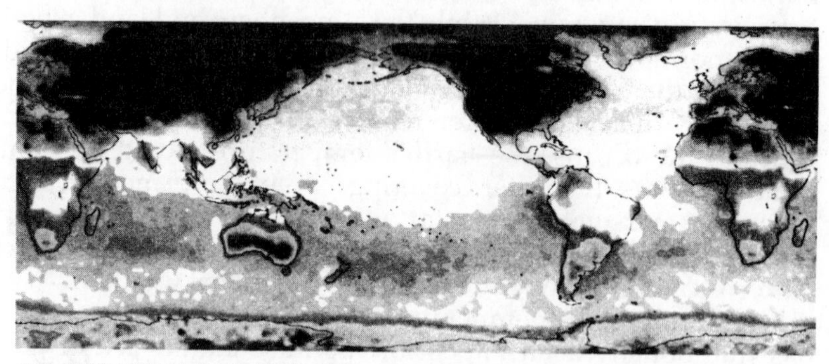

TEMPERATURE DIFFERENCES

Fig. 2.10. Global surface temperatures (1).

data from the High Resolution Infrared Sounder and the Microwave Sounding Unit. Both of these instruments measure natural radiation emitted from the Earth's surface and atmosphere, and have been flying on NOAA weather satellites since 1979. Temperatures below freezing (273°K) are grey on top and bottom. Warmer temperatures are white and grey (centre).

(Top frame in Fig. 2.10). In January 1979 the Northern Hemisphere is experiencing extreme cold. Siberia and most of Canada record temperatures approaching -30°C. In Eastern Europe and the northern U.S.A; temperatures are below 0°C. In the Southern Hemisphere, the people at mid-latitudes (30-50°) are enjoying summer with temperatures ranging from 20-30°C. In the open oceans, the isotherms (or contours of equal temperature) show deviations from their zonal patterns on the eastern and western sides of the various oceans. Generally in the subtropics of both hemispheres (10-30°), the western sides of the oceans are warmer than their eastern counterparts, primarily due to ocean currents. An exception to this rule is the Gulf Stream, a warm current in the western Atlantic. The current moves along the North American continent, then turns north-east-ward transporting warm waters across the Atlantic that moderate the climate of Western Europe.

(Middle). By July, areas of the Northern Hemisphere have warmed to 10-20°C. Equatorial Africa and India are the hottest in dramatic contrast to the frozen Himalayan Mountains. In the Arctic, Greenland remains frozen, while Hudson Bay has thawed.

In the Southern Hemisphere, Antarctica is much cooler than the Arctic and ice has formed in the Weddell Sea. At this high latitude, zones of constant tempertature are much more zonal than in the northern oceans. Here, the ocean driven by strong westerly winds moves in a circular path around Antarctica from west to east.

(Bottom). Temperature differences between January and July show that the greatest warming and cooling has occurred over land (black). Marked seasonal changes of up to 30°C are seen in both hemispheres. In contrast the changes in ocean temperature rarely exceed 8-10°C. The greatest deviations are in the mid-latitudes, while the near-equatorial regions are quite stable. In the Northern Hemisphere, mid-latitude changes in ocean temperature are influenced by the position and continents. The continents divert the ocean currents and affect wind patterns. In the Southern Hemisphere, which has one-half the land area of the north, changes are primarily due to seasonal variations in incoming solar radiation. Thus, the oceans in the two hemispheres interact in fundamentally different ways with the atmosphere and land.

Another area of investigations is the determination of snow-cover by remote sensing. Figures 2.11 shows an example of a daily snow map for eastern Europe from Scanning Multichannel Microwave Radiometer data for October 31, 1978 (NIMBUS-7). Of special interest is the monitoring of the ice cover of the poles (see Fig. 2.12)

Ice and snow cover a significant portion of the Earth's surface and thus strongly influence global climate. In winter, Antarctic sea ice covers an area greater than that of the United States. Driven by winds and ocean currents, most of this ice is constantly moving. It acts as an insulating barrier that regulates the rate at which heat can be transferred between the atmosphere and oceans. This has a major but poorly under-

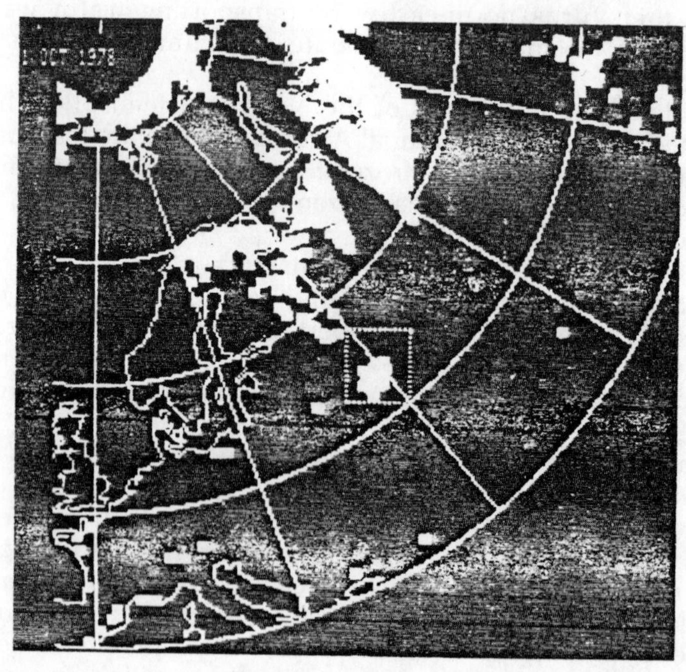

Fig. 2.11. Snow map (2).

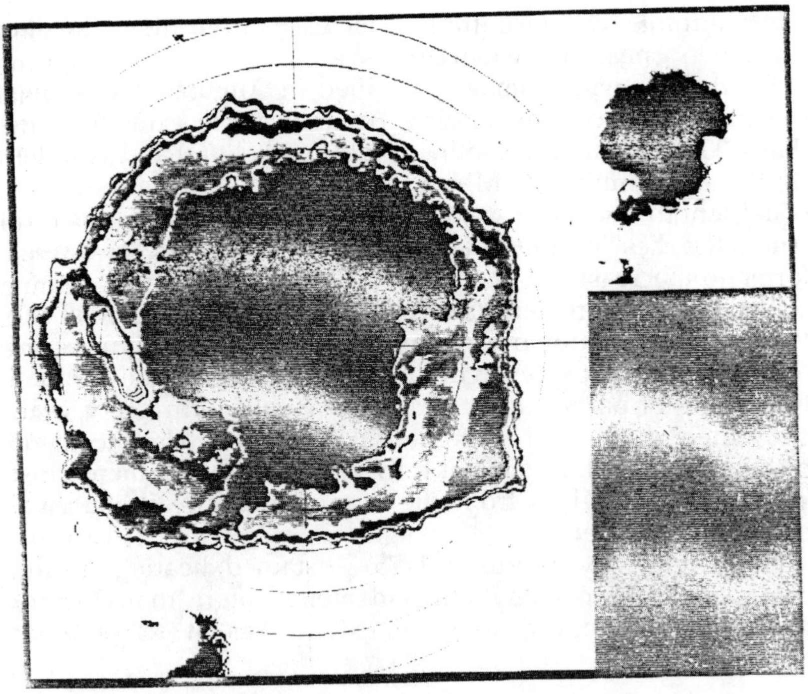

Fig. 2.12. Antarctic sea-ice (1).

stood effect on global climate. Mapping seasonal and year-to-year variations in the sea-ice cover is critical in developing the understanding of climate.

Until the advent of satellite observations, annual and year-to-year changes could not be accurately measured. Early weather satellites provided visible imagery of the sea-ice cover. There are distinct differences between the intensity of microwave emissions from water and from sea-ice, so the two can readily be distinguished. In areas where there is a mixture of ice floes and open water, sea-ice concentration can be estimated. Microwave sensors have been in orbit since 1973, scanning the entire globe every few days.

The images in Fig. 2.12 show patterns of seasonal freezing and melting of sea-ice around Antarctica. They were obtained by analysing microwave-emmission measurements from the Electrically Scanning Microwave Radiometer (ESMR) aboard NASA's NIMBUS-5 satellite. The Antarctic land mass is overlaid in black, with the tip of South America in the upper left quadrant. Ice concentration decreases from high near the coast to low near the ice margin.

(Left). This winter image, obtained in August 1974, shows the full ice cover surrounding Antarctica. Ice extends more than 1000 km from the continent into the Ross (1) and Weddell (2) Seas. Beginning in March, when days and nights are of equal length, ice-cover gradually increases. By May, much of Antarctica lies in continual darkness and the rate of sea-ice formation increases.

The large, ice-free enclosure seen in the eastern part of the Weddell Sea is known as a polynya. Although polynyas are found in many areas, even in the depths of winter, the Weddell Polynya is of particular interest. It does not form every year, but when it does, it is found in approximately the same position. Why it forms is not fully understood, but there must be a major heat loss from the ocean to ensure its survival through the winter.

(Top right). By February 1975, sustained heating during continuous summer daylight had melted more than 80% of the winter ice cover, and ice extent was at its summer minimum.

5.4. Synthetic Aperture Radar (SAR)

A spaceborne remote sensor that gains more and more interest is the Synthetic Aperture Radar. The more data processing techniques and technologies improve, the more SAR has proven itself as a viable tool for determination of global sea, ice and land characteristics. Independent from sun illumination, clouds and most of the weather conditions, it provides clear images of the ground scenes over a wide swath and with high spatial resolution. One of the most famous SEASAT SAR images is shown in Fig. 2.13 showing the area of Los Angeles. This image was processed by the Jet Propulsion Laboratories (JPL). It shows the flat town area (centre) with streets (black lines) and buildings (white spots), some farmland (areas of different grey levels with typical contours), the mountains (top), the Pacific coast line (bottom left), ships near the coast (white dots) and some strange features on the sea being caused by specific ocean wave structures.

Fig. 2.13. Los Angeles by SAR.

The next example for the use of SAR images is given in Fig. 2.14. A sketch of water bodies and drainage features was derived from SEASAT SAR imagery over Florida (Bryan, 1981). In this case the general advantage of remote sensing in providing land-cover data for water resources investigations is that the sensor system can provide parameter estimates for individual hydrologic-response units over geographically large areas in a cost-effective manner.

The next SAR Image shown (Fig. 2.15), 40 km wide, was obtained by NASA's SEASAT in September 1978, during its pass over the Upper Chesapeake Bay, Maryland. Shoreline detail is clear, as is the Chesapeake Bay Bridge (1). Water along many of the east-facing shorelines appears dark. This indicates smooth surface waters in a sheltered position leeward to the wind. Streaks (2) above the Bay Bridge and lighter coloured water at the bottom left result from roughening of the water surface by bay currents and by wind, respectively. South of the

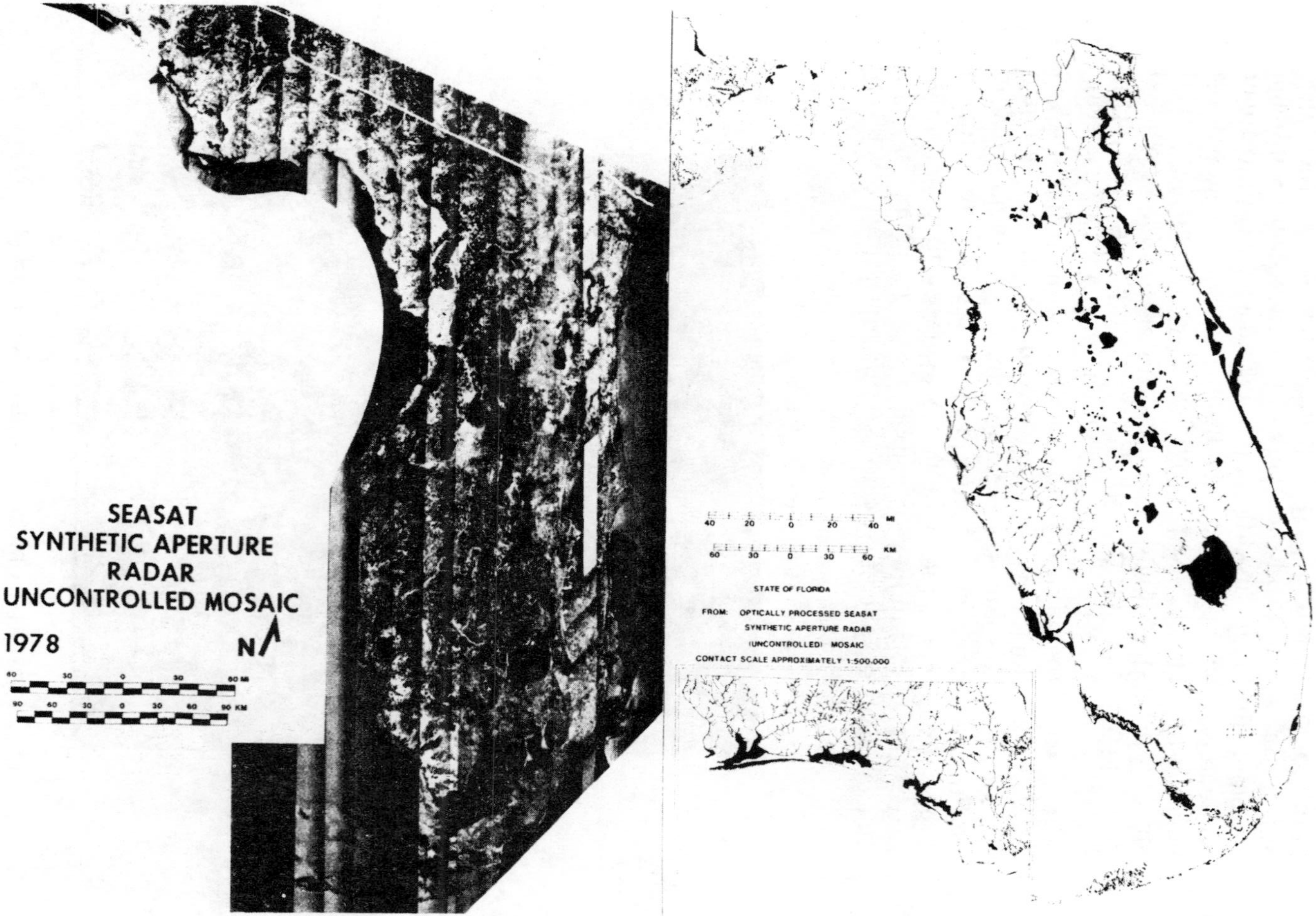
SEASAT
SYNTHETIC APERTURE
RADAR
UNCONTROLLED MOSAIC
1978
N
60 30 0 30 60 MI
90 60 30 0 30 60 90 KM
40 20 0 20 40 MI
60 30 0 30 60 KM
STATE OF FLORIDA
FROM: OPTICALLY PROCESSED SEASAT
SYNTHETIC APERTURE RADAR
(UNCONTROLLED) MOSAIC
CONTACT SCALE APPROXIMATELY 1:500,000

Fig. 2.14. SAR image of Florida and the derived sketch of water bodies and drainage features (3).

Fig. 2.15. Upper Chesapeake Bay (SAR) (1).

bridge, three large anchored ships (3) are seen as white spots. To the north are moving ships with their trailing wakes (4). The thin dark line heading west from the bridge and crossing an inlet is U.S. Route 50. Near the inlet mouth (centre left) Annapolis appears as clustered white spots. Buildings in cities produce a bright return from their many angular reflecting surfaces.

SAR images also reveal other aspects of sea surface structure, including shallow water shoals, surface and internal waves, oil slicks and polar sea ice. These detailed images are adding to a gallery of SAR pictures that have both scientific and practical value.

As mineral and petroleum exploration and exploitation in the Arctic continue, safer and shorter sea routes are being sought. Movement of sea-ice poses a serious hazard to activities in the region. Satellite imagery has demonstrated the great potential for monitoring ice formation and movement in the polar oceans (Fig. 2.16).

A last introduction to SAR imagery is given by Figs. 2.17–2.19, showing special land structures of different areas in the world.

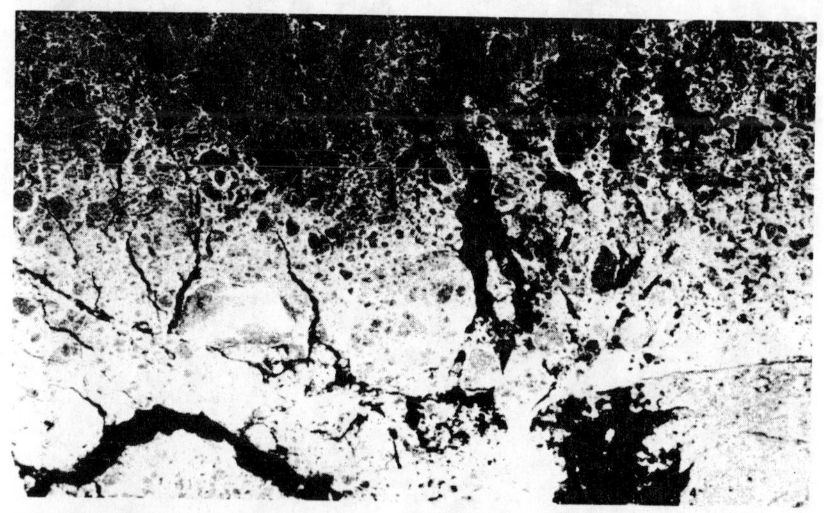

Fig. 2.16. Arctic ice (FAR) (1).

6. OUTLOOK

Radar remote sensing from space plays an increasingly important role in the global earth scientific research programme concerning meteorology, oceanography and land applications. Earth science is already faced with problems that are truly global in extent and interdisciplinary and multidisciplinary in nature. Since the land, ocean and atmosphere are coupled, many of the foremost questions can no longer be treated in isolation, but require observations of the system as a whole. Progress can be made piecemeal, but the interdisciplinary questions require observations and effort across the broad spectrum of Earth science. An important example is the study of biogeochemical cycles, the processes by which key chemical

Fig. 2.17 (left). Cultivation fields in central New South Wales (Australia) (SAR) (4).
Fig. 2.18 (right). Circular fields in south-western Libya (SAR) (4).

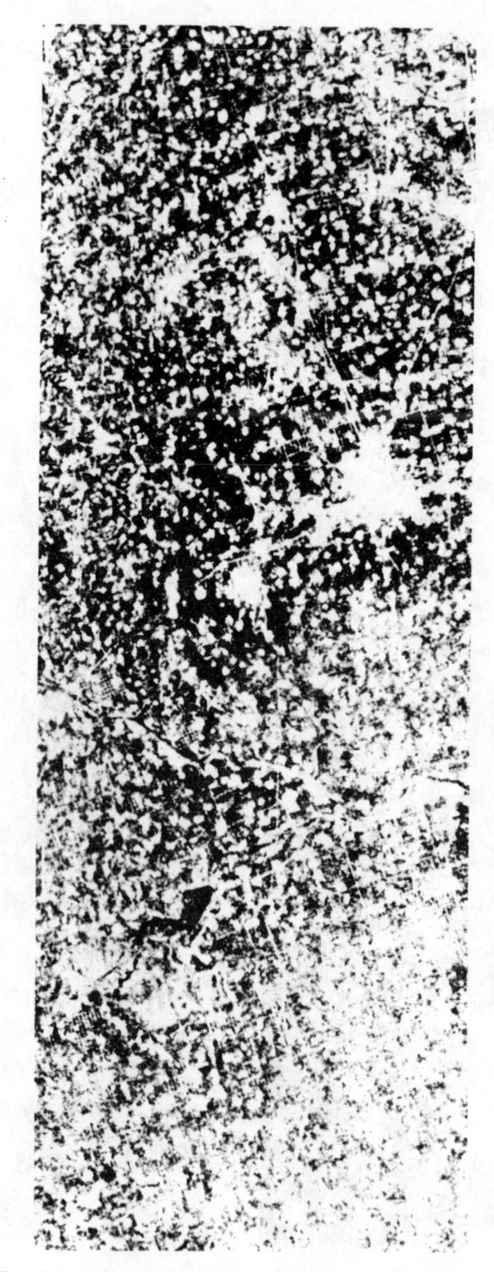

Fig. 2.19. Land use in northern China (SAR) (4).

elements are transformed and exchanged among the soils, biota inland waters, oceans and atmosphere. A second example is the hydrologic cycle, in which water is cycled among the oceans, atmosphere, cryosphere and land surface. While these are distinct problems, they are also important parts of the study of climate in which the interactions among atmospheric composition, soil moisture, surface albedo, cloud amount, ocean heat transports and atmospheric dynamics couple together to determine the environment in which life takes place. These global problems are vital to all the inhabitants of our planet; progress towards solving them will require the creative involvement of the international scientific community.

Beside these global investigations monitoring of natural resources is a major aim of spaceborne remote sensing. Optical images are combined with SAR images over land and coastal zones and auxiliary information is used concerning weather, coastal ocean features and temperatures being derived by altimeters, scatterometers and radiometers etc. The work of many national centres of remote sensing during the last decade has fruitfully supported these ideas and has helped to understand some of the strange features being detected in radar images. Nevertheless, an unambiguous detection of special natural resources needs about one more decade of radar signature research, the development of complex sensor types and extensive sensor calibration methods, and last but not least it needs operational remote sensing programmes providing a reservoir of data from different sensor types for the user to experiment with.

7. REFERENCES

(1) 'Oceanography from Space'. NASA Publication.

(2) Kunzi, K.F., Patil, S. and Rott, H. Snow-cover parameters retrieved from Nimbus-7 SMMR Data. *IEEE Trans.*, 4 Oct. 1982.

(3) American Society of Photogrammetry. *Manual of Remote Sensing*. The Sheridan Press.

(4) *Proc. IGARSS' 82*. Munich, IEEE.

(5) *Proc. IGARSS' 84*. Strasbourg, IEEE.

(6) *Science Program for an Imaging Radar Receiving Station in Alaska*. NASA/JPL publication, 1983.

(7) Technical Memorandum 86 129. *Earth Observing System*. NASA/Goddard Space Flight Center.

(8) *Antarctic Sea Ice, 1973-1976 Satellite Passive-Microwave Observations* NASA SP-459.

3

SPOT: The First Operational Remote Sensing Satellite

*Gérard Brachet**

ABSTRACT

The first SPOT satellite will be launched in October, 1985. SPOT 2, identical to SPOT 1, will be available for launch in early 1987. They will open a new era in land remote sensing with a clear operational objective featuring a worldwide commercial distribution network.

The SPOT satellites will produce images of the earth's surface at 10 m sampling intervals in the panchromatic mode, and at 20 m sampling intervals in the multispectral mode. Imaging can be made either along the vertical, or off-nadir to provide repetitive coverage or stereo pairs for altitude determination. In addition, a variety of image preprocessing and processing levels will be available according to the user needs.

The many innovative features of the SPOT system, combined with an efficient distribution organization, responsive to market requests, should provide the user community with high value information in a wide range of application areas. Preliminary results have already been obtained from the many SPOT simulation data collected in Europe, West Africa, Bangladesh and the United States. They confirmed that in such fields as topographic and thematic mapping, forest inventory, crop production statistics, urban planning, engineering and geology, the SPOT images may very well replace high altitude aerial photography.

Combining a lower price on a square kilometre basis and a higher information content (radiometric accuracy), ready for digital processing, they will gradually be accepted by the user community in these various areas, provided that a clear prospect of system coninuity is available.

* Address of the author: Dr Gérard Brachet, SPOT IMAGE, Toulouse, France.

1. INTRODUCTION

The SPOT programme has been decided and designed as an operational and commercial system. Decided by the French government in 1978, with the participation of Sweden and Belgium, the programme is managed by the French Space Agency (CNES) which is responsible for the programme development and satellite operations. SPOT 1 will be launched in October 1985 and SPOT 2, to be available for launch in early 1987 as a back-up for SPOT 1, is also under construction. Plans are being made for the launch of SPOT 3 and 4 in 1990 onwards in order to ensure the necessary service continuity expected from an operational spaceborne remote sensing programme. Actually, it is essential that current programmes be followed up over a sufficiently long period (at least 10 years) to allow the development of applications in those areas where remote sensing is not yet widely used. This is also necessary to implement adequate procedures for operational aspects, involving training activities and equipment.

The organizational structure adopted for the management of the SPOT programme clearly distinguishes between the functions of technical system management, executed by CNES and the responsibility, assigned to SPOT IMAGE, a commercial corporation, for relations with the user community and the data distribution. This structure should ensure the efficient management of both satellite image acquisition capacity and data transmission, by the ground network of SPOT data-receiving stations, in accordance with user requirements. In addition, the conditions of image distribution on a commercial basis go hand-in-hand with the necessity towards the complete auto-financing of the programme. This policy should make it possible to ensure continuity of service beyond the first two satellites.

2. BRIEF RECALL OF THE SPACECRAFT CHARACTERISTICS

The SPOT spacecraft carries two identical sensors, called HRV (Haute Résolution Visible), made of static solid state arrays of detectors (CCD) and operating in the visible and near infrared part of the spectrum. Among the innovative features of SPOT are the relatively high ground resolution of the imagery it will produce (10m in the panchromatic mode, 20m in the multispectral mode) and the ability of its sensors to point up to 27° east and west of the local vertical axis. This latter feature offers interesting possibilities for increasing the number of opportunities to obtain views of a given area. It also permits stereoscopic observations by combining views taken at different angles from the vertical and therefore opens up the possibility of third dimension (or altitude) determination, an important requirement for cartographic applications. The principal characteristics of SPOT are summarized in Table 3. I.

Table 3.1 SPOT : Principal characteristics.

Orbit	circular at 832 km inclination : 98.7 degrees descending node at 10h 30mn a.m. orbital cycle: 26 days
Haute resolution visible (HRV)	two identical instruments pointing capability: ± 27° east or west of the Orbital plane ground swath: 60 km each at vertical incidence pixel size: 10 m in panchromatic mode 20 m in multispectral mode spectral channels: panchromatic: 0.51 — 0.73 μm multispectral: 0.50 — 0.59 μm 0.61 — 0.68 μm 0.79 — 0.89 μm
Images transmission	two on board recorders with 23 min capacity each direct broadcast at 8 GHz (50 Mbits/s)
Weight	1750 kg
Size	2 × 2 × 3.5 m plus solar panel (9 m)

2.1. Swathwidth

Two identical sensors (HRV) are on-board the satellite and can be activated independently. Each instrument has a swathwidth of 60 km. When the two instruments operate in adjacent covering field, the ground coverage is 117 km.

2.2. Imaging Modes

SPOT operates in two modes: multispectral mode and panchromatic mode. In the multispectral mode, observations are made in three spectral bands with a pixel size of 20 m:

- a green band from 0.50 μm to 0.59 μm;
- a red band from 0.61 μm to 0.68 μm;
- a near infrared band from 0.79 μm to 0.89 μm.

In the panchromatic mode, observations are made in a single broad band, from 0.51 μm to 0.73 μm with a pixel size of 10 m. The multispectral bands have been selected to take advantage of interpretation methods developed over the last ten years; they have been designed to allow the best discrimination among crop species and among different types of vegetation using three channels only.

The panchromatic band will offer the best geometric resolution (10 m) and will make it possible to comply with cartographic standards for maps at a scale of 1:1 000 000 and/or to update at a scale of 1:50 000 and in some cases 1:25 000 for thematic applications.

2.3. Field pointing flexibility, Nadir and off-nadir viewing

One of the key features of SPOT is the steerable mirror which provides off-nadir viewing capability. The instrument can be tilted sidewards (to the east or to the west) step by step from 0 to 27° allowing scene centres to be targeted anywhere within a 950 km wide strip centred on the satellite track. This technique provides a quick revisit capability on specific sites. For instance, at the Equator, the same area can be targeted seven times during the 26 days of an orbital cycle i.e. 98 times in one year with an average revisit of 3.7 days. At latitude 45°, the same area can be targeted 11 times in a cycle i.e. 157 times in one year, with an average of 2.4 days, a maximum time-lapse of 4 days and a minumum time-lapse of 1 day. (Fig. 3.1).

The revisit flexibility allows:

- monitoring of phenomena which rapidly vary over time, such as crops, environmental stresses, natural disasters;

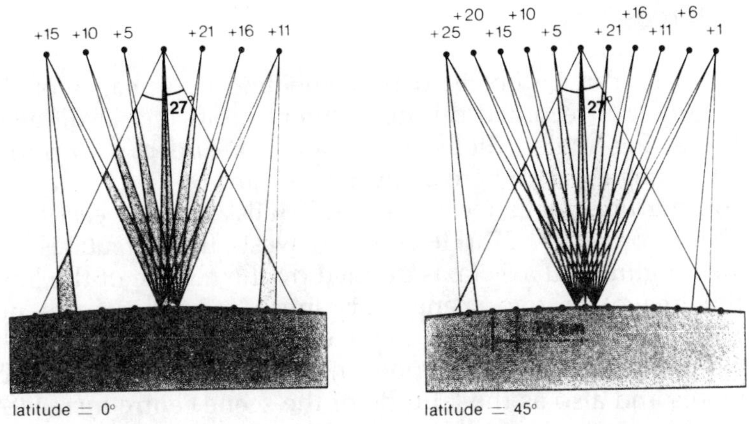

Fig. 3.1. Typical sequence of acquisitional at the Equator and at latitude 45°.

- improving the possibility of obtaining timely data required in many studies;
- improving the rate of coverage by minimizing the effects of weather conditions.

Figure 3.2 illustrates the variability in obtaining a complete coverage of France using cross-track viewing capability. At vertical viewing the coverage is obtained in 313 days, and in 100 days if 27° depointing mode is used.
Figure 2: Time required to obtain a complete coverage of France for different maximum cross-track viewing capabilities.

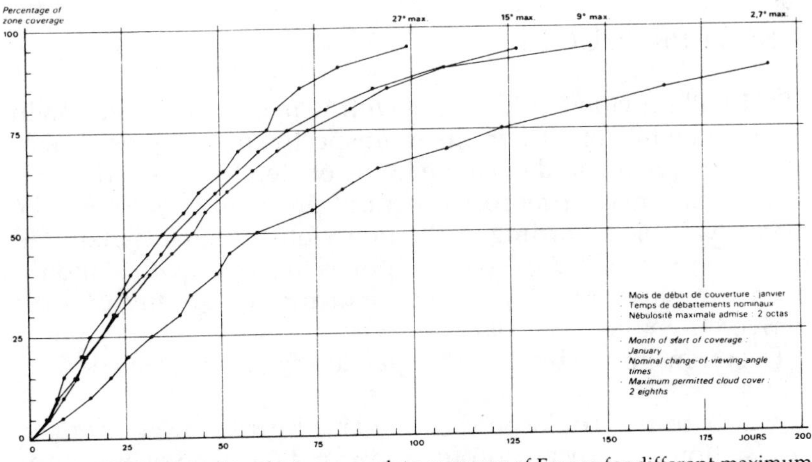

Fig. 3.2. Time required to obtain a complete coverage of France for different maximum cross-track viewing capabilities.

2.4. Stereoscopy

Stereo pairs can be obtained by combining two images of the same zone but recorded during different orbits and at different HRV viewing angles. Stereo pairs can be recorded with mirror pointing angles anywhere between -27 and +27°. For a pair comprising one image recorded at -27° (HRV looking east), and another recorded at +27° (HRV looking west), the B/H ratio is 1.

The longitudinal axis (axis defined by the middle of the lines of the image) of two scenes forming a stereo pair are not parallel. The angle between the axes of two scenes forming a stereo pair increases as the pointing angle of the HRV mirror increases and also as the latitude of the scene centre increases. Stereopairs are an important requirement in many cartographic applications: geomorphical, geological or soil maps and of course for topographic maps. SPOT will provide the opportunity to map anywhere in the world with a mapping accuracy corresponding to the 1:100 000 standards.

2.5. Images transmission

Direct broadcasting operates at 8 GHz at a rate of 50 Mbits/s. The satellite carries two on-board recorders with a 23 min capacity each. On-board data recording will be used over areas where no ground receiving facility occurs.

3. SPOT PRODUCTS

SPOT will provide Earth images with 10 m ground resolution in black and white or 20 m in the multispectral mode, at vertical or off-nadir viewing and in single frame or stereopair. In addition, a variety of image preprocessing and processing levels will be made available according to the precision and quality required by users for 'SPOT data' which applies to scenes with standard processing and 'SPOT products' which refers to any product derived from the above data.

Data transmitted by the SPOT satellites will be received:

- at both the Toulouse (France) and Kiruna (Sweden) main stations in direct read-out mode for data taken over Europe and Polar zones. These two stations will also receive the worldwide data dumped from the onboard recorders. Each of them has a receiving capacity of 250 000 scenes per year;

The preprocessing centres attached to both stations and operating in Sweden and in France, have a capacity of 70 system corrected (level 1) scenes per day or 20 precision processed (level 2) scenes per day. A level 1 scene can be preprocessed within 48 hours from its acquisition at the ground station in standard procedure, when a precision processed scene requires 5-7 days;

- at other receiving stations in the world which will receive only real time data over the stations visibility range.

The distribution of SPOT data and products will be made by the receiving stations within the station's distribution zone which generally correspond to the country or countries operating the station. Elsewhere the distribution will be ensured by SPOT IMAGE and a network of national distributors. Therefore, users will have various possibilities for information purposes or for product ordering:

- to contact the national receiving station acting as a distributor when they are resident in such a country;
- to contact the national distributor(s) in those countries that do not possess a receiving station;
- or to contact directly SPOT IMAGE in Toulouse.

Information availibility through easy and quick access to data and products is a very important factor in operational procedures. The system implemented for data acquisition as well as data distribution has been conceived to meet the user needs and SPOT customers are given an active part in the system.

The basic unit for segmenting the image data streams at the ground receiving and preprocessing stations will be the 'scene' which corresponds to the totality of image data for an area 60 km in length (along the ground track) and 60–80 m in width (in the cross-track direction) depending on the instrument viewing angle. This corresponds, for vertical (nadir) viewing to 6000 × 6000 pixels per scene in the panchromatic mode and 3000 × 3000 pixels in the multispectral mode (Table 3.2).

In the case of cross-track viewing (i.e. for mirror pointing angles between 0° and the maximum value of 27°), the length of a scene in the ground track direction is always 60 km, but the

Table 3.2 Basic characteristics of spot scenes

	XS mode	P mode
Scene dimensions (nadir viewing): 60 × 60 km		
Pixel size:	20 × 20 m	10 × 10 m
Number of spectral bands:	3	1
Dimensions of preprocessed scenes:		
number of pixels per line: (raw scene to level 2 scene)	3 × (3000 — 5200)	6000 — 10400
number of lines per scene: (raw scene to level 2 scene)	3 × (3000 — 4900)	6000 — 9800
volume (8-bit bytes):	27 — 76.5 Mb	36 — 100 Mb

width, in the cross-track direction, is a function of the mirror pointing angle and ranges between 60 and 80 km. When the two HRVs are operated in the so-called 'twin mode' with the mirror pointing angles near 0°, the result is termed a 'bi-scene'. Such SPOT bi-scenes cover 60 × 117 km with 3 km overlap between the two.

SPOT IMAGE and stations operators will deliver basic image data preprocessed in a number of different ways to make them usable. This applies to radiometric corrections taking into account the calibration factors for the detectors and the optical and telemetry systems and to geometric corrections to take account of the viewing conditions and of the precision required for specific applications. Four main levels of preprocessing are anticipated roughly corresponding to system corrected (level 1B) and precision processed (level 2 and S) data.

3.1. Level 1A

This is essentially 'raw' data, the only processing performed being the equalization of the response of the CCD detectors. Neither interband calibration nor geometric correction is applied. Level 1A data are intended for users requiring imagery that has undergone a minimum of preprocessing and in partiular for stereo plotting purposes.

3.2. Level 1B

This level involves radiometric and geometric system corrections [compensation of: rotation of the earth, satellite perspective effects, viewing angle and effects of satellite forward motion (desmearing)]. The location accuracy is 1500 m (rms), for vertical viewing, and the relative error 10^{-2}. This is the basic preprocessing level for photointerpretation and thematic analysis. Stereoscopic pairs, with different B/H ratios depending on instrument viewing angles, are available at this level.

3.3. Level 2

This is a precision processed level. Radiometric correction is as for level 1B. Geometric corrections involve bi-dimensional computation based on 6-9 ground control points (GCPs) per scene. The image is rectified according to a given cartographic projection: Lambert Conformal, Transverse Mercator, Oblique Equatorial, Polar Stereographic or Polyconic. The location accuracy is 50 m (rms) for vertical viewing. But this level does not take account of disortions due to relief. Thus, the closer the viewing angle to the vertical and the less pronounced the relief, the more accurate the final product. Film reproductions of level 2 data are oriented with geographic North corresponding to the 'Y-axis' of the film.

3.4. Level S

This level of preprocessing involves scene rectification relative to landmarks to ensure registration with another scene used as a reference to within 0.5 pixel, i.e. to within 5 or 10 m depending on the imaging mode. Level S products are typically used in multidate studies.

These basic standard preprocessing levels are performed by the preprocessing centres at Toulouse (CNES) or Kiruna (SATIMAGE) or at other preprocessing centres attached to local receiving stations.

In addition to these basic preprocessing levels, a number of further processed and value added products will be made available. These products include most of the multispectral enhancement and processing for cartographic purposes:

- cartographic precision processed SPOT scenes: level 3 (orthophotography using a DMT) and level 4 (production of contour level maps) such as those that will be produced by Institut Géographique National;

- radiometric enhancements: stretching, edge enhancement, etc.;
- channel combinations: ratios, PCA, colour composites of basic or new computed channels, etc.;
- mixed products such as SPOT panchromatic and multispectral scene or SPOT scenes merged with other sources of information;
- geocoded products;
- multispectral classifications.

SPOT products will be available, by full scene or sub-scene, recorded on the following media: CCT, floppy disk, photographic film or print.

3.5. Magnetic Media

3.5.1. Computer Compatible Tapes (CCTs)

The SPOT CCT format belongs to the 'CCT Family of Tape Formats' as defined by the Landsat Ground Station Operators Working Group (LGSOWG). SPOT CCTs may be recorded at 6250 to 1600 bits per inch (bpi). Recall that a single SPOT scene comprises from 27 to 100 million bytes depending on the imaging mode and the processing. This means that one full SPOT scene can be recorded on one 6250-bpi tape or on two or three 1600-bpi tapes each with a maximum capacity of 32 M bytes. The image record length will be, depending on the case, 5400, 8640 or 10890 bytes (of 8 bits). The recording format will be band-interleaved-by-line (BIL) (Table 3.3).

Table 3.3. Basic characteristics of SPOT scenes.

Each 60 × 60 km SPOT scene will be available:
as panchromatic data (one band, 10 m)
or as multispectral data (three bands, 20 m)
at 4 processing levels:
1a equalization of detector response no geometric correction
1b radiometric correction and geometric correction induced by the acquisition system
2 radiometric correction plus precision geometric correction to map the image into a cartographic projection
S registration with a reference scene
in single frame (at Nadir or off-Nadir viewing) or in stereopair
on photographic film at scales ranging from 1/400 000 to 1/25 000 or on CCT (1600 or 6250 bpi)

A tape will contain five files: a volume directory, leader file, imagery data file, trailer file and null volume directory. The leader file contains ancillary information concerning the satellite, the recorded scene and preprocessing (ephemeris data, attitude; calibration coefficients and previously computed histograms, etc.).

Radiometric levels will be encoded using 254 levels. Level O corresponds to non-significant values (borders, losses of synchronization) and level 255 is reserved for future applications.

3.5.2. Floppy disks

Each floppy disk will only be able to store a small part of a scene. The format for the recording of floppy disks has yet to be determined.

3.6 Photographic Media

The basic photographic medium is the 241 × 241 mm format film corresponding to a full SPOT scene preprocessed to level 1B and at a scale of 1:400 000. However, in view of the very fine detail in such imagery, it is not directly usable at this scale, i.e. it needs to be enlarged. The basic film format for level 2 products will be 350 × 350 mm. In this case, North-oriented images will require yet another format because of their 'skewness'. Level 1B, 2 and S film products will be available in black and white (panchromatic or three spectral bands), or as colour composites, with the following scales and formats:

Level	Film format	Scale	Content
1B	24 × 24 cm	1 : 400 000	1 full, 60 × 60 km scene
	24 × 24 cm	1 : 200 000	¼ scene
	50 × 50 cm	1 : 200 000	1 full scene
	50 × 50 cm	1 : 100 000	¼ scene
	50 × 80 cm	1 : 100 000	½ scene
2	35 × 35 cm	1 : 400 000	1 full, 60 × 60 km, scene
	35 × 35 cm	1 : 200 000	¼ scene
	70 × 70 cm	1 : 200 000	1 scene
	70 × 70 cm	1 : 100 000	¼ scene

Level 3 film products will be available in two formats: 50 × 50 cm and 100 × 100cm.

Paper prints will be available with scales starting at 1:200 000. Level 1 quarter-scenes will use the 24 × 24 cm, and level 2 the 35 × 35 cm format. Quarter and full scenes will also be available at 1:100 000 using correspondingly larger format.

3.7 Quick-look Images

Quick-look images can be briefly characterized as follows:

- gross geometric and radiometric correction;
- panchromatic mode data subsampled at the rate of 1 pixel in 6 and 1 line in 6;
- multispectral mode data subsampled at the rate of 1 pixel in 3 and 1 line in 3.

Quick looks will be available as photographic paper prints at a scale of 1:400 000 (format: 24 × 24 cm), unless ordered on a subscription basis, in which case they can be recorded on videocassettes.

4. POTENTIAL APPLICATIONS

In 1979, it was decided as part of the preparation for and the promotion of the SPOT programme, to conduct a series of campaigns aimed at simulating the types of imagery that will be recorded by the SPOT earth observation satellites. At first this exercise was simply intended to meet the requirements of users who had participated in the drawing up of the specifications for the SPOT system and to ensure the compatibility of SPOT products with the intended applications in such areas as cartography, agriculture, geology, land use, the environment, and so on.

In view of the interest created by the first simulated images, this programme, which was open only to French laboratories during the first year, was quickly extended to cover a number of foreign countries and made open to all users. Thus, all potential users had (and still have) an opportunity to become acquainted with simulated SPOT data, to assess the technical qualities of the products, and to evaluate the usefulness of the data for different applications.

To date, 400 simulated scenes, covering 23000 km^2 and 130 sites, including a wide variety of geographical zones, have been prepared. Some 50 or so organizations and laboratories have participated financially or technically in the campaigns.

The first analyses completed have yielded promising results suggesting that SPOT data should be well suited to a wide range of applications from which some examples have been selected to illustrate this paper.

4.1. Land Use and Agriculture

With regard to land use studies in rural environments, it appears certain that the high resolution of SPOT images will prove to be a major advantage.

An interesting case in point is the visual interpretation study conducted by the Bureau pour le Dévelopment de la Production Agricole (B.D.P.A.) of several areas of Corsica. This work revealed that, without doubt, SPOT imagery is particularly suitable for thematic mapping in highly complex regions such as the Mediterranean basin and similar regions in other parts of the world, characterized by a complex pattern of landscapes.

In these areas, currently available satellite remote sensing data has been of little use due to insufficient resolution.

The potential usefulness of SPOT data for such applications is due to the judicious choice of spectral bands which, associated with high resolution (10 m) of the panchromatic mode, greatly facilitates the identification and location of major features, despite their frequently small dimensions. The excellent accuracy of feature location also greatly facilitates the use of ground truth data and the mapping of the results. With regard to mapping, it should be noted that the use of standard level 2 SPOT products will completely bypass the problem as these will conform to cartographic projections.

Given the richness of textural and structural information contained in the simulated data, SPOT imagery would appear to be fairly comparable to small-scale aerial photography (1:60 000 to 1:100 000).

Analysis of the results obtained for the types of terrain studied reveals that the 20 m resolution multispectral and 10 m resolution panchromatic modes are complementary and that the corresponding data should thus be used in association or in combination : panchromatic data are nearly always required to bring out textural and structural features, while multispectral data provide complementary information on vegetation, soil and rock types, water bodies etc.

In image digital processing and automatic classification, the pixel size compared with field, parcel or plot size is of capital importance.

With Landsat MSS 80 m resolution, a field needs to be 2.56 ha in area before it surely contains one pure pixel . . . and this for a total of 8 overlapping border pixels. With 20 m resolution, the corresponding area is 0.16 ha while with 10 m resolution, it is 0.04 ha. If we consider the case of a field of 2.5 ha and 20 m resolution imagery, we find, on average, 49 pure pixels (i.e. fully within the parcel) and 15 border mixed pixels. Thus, the ratio of pure to mixed pixels is approximately 4 to 1. This means that it should be possible to unambiguously distinguish between the two in virtually every case.

Since 1981, the Centre National d'Etudes Spatiales (C.N.E.S.) and the Service des Enquêtes et Etudes Statistiques (S.C.E.E.S.) of the French Ministry for Agriculture have been using simulated SPOT data of a number of test sites in the Laurangais region of SW France (approximately 50 km S.E. of Toulouse), to study the possibility of incorporating remote sensing data into the agricultural statistics system. Each year this system estimates the areas of land under different crops, the data being derived from the annual TERUTI statistical survey which, in turn, is based on aerial photographs and ground truth. In addition, there is a general agricultural census (RGA) once every ten years. This is a much larger operation as it takes into account various socio-economic parameters.

The Lauragais area can be considered as representative of many parts of Europe as regards the average field size (2 ha) and the wide variety of crops grown : wheat, corn, sorghum, rape and sunflower, not to mention interspersed uncultivated fields and woods. Despite the technical difficulties associated with this simulation exercise, and particularly with the registration of airborne scanner data acquired on four different dates, the results obtained are significant.

They reveal, for instance, that data acquired on two dates (May 6 and June 30) can result in a better than 90% correlation between pixels classified according to crop type (cartographic quality) and ground truth data, and that the error associated with the estimation of the corresponding crop areas (overall ratio of estimated area to actual area) is less than 5%. (Fig. 3.3).

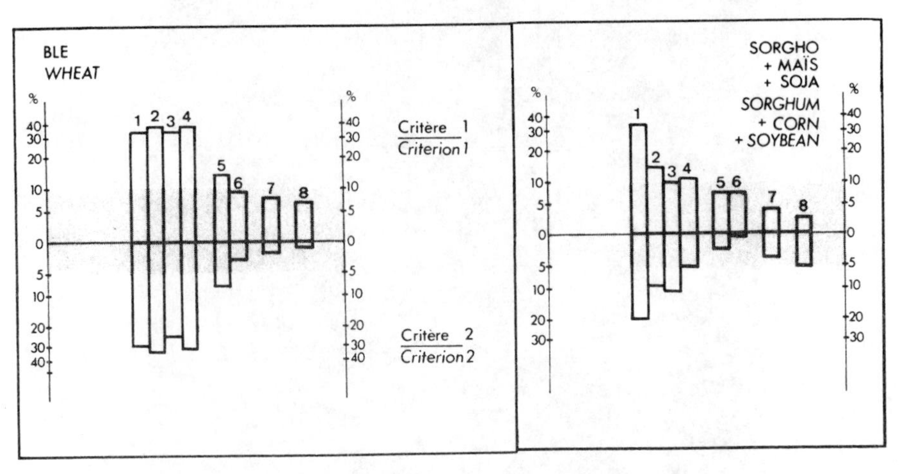

Fig. 3.3. Results of classification.

This study has also demonstrated the importance of judicious choice of acquisition dates relative to the different crop cycles. The difficulties associated with unreliable weather at that time of year (May-June) in Western Europe means that users requiring data collection at such periods should draw benefit from the inherent flexibility of the SPOT system as regards the revisit capabilities. In this connection, for instance, a change of just 5° pointing angle of the swatch selection mirror of the HRV instrument results in the doubling of the acquisition capacity over a given area.

4.2. Urban and Peri Urban Planning

The efficient planning and management of urban areas can only be achieved through access to accurate and continuously updated information concerning land-use and changes therein. Today, aerial photographs are still the prime source of such data.

Naturally, the high-resolution is appreciated, but aerial surveying is an expensive exercise, particularly when it comes to the analysis of multidate photos. Attempts to use 80 m and even 30 m resolution space imagery in this area have only yielded modest results, as the resolution is not comparable with the typical sizes of urban objects, particularly in the case of old pattern cities with complex structures. The simulated

Fig. 3.4. simulated image of Paris West.

SPOT imagery of such cities as Paris and Toulouse in France, or Dakar, Senegal, and Ouagadougou, Upper Volta, in West Africa reveal that 10m resolution is about sufficient for the identification of the main features of a typical urban pattern. For instance, comparison of the simulated SPOT imagery of Paris with maps at different scales prepared by the Institut d'Aménagement et d'Urbanisme de la Région Ile de France (IAURIF) reveals that the images are richer in detail and more accurate than 1:100000 maps and that there is perfect correlation, as regards the road network, between the images and 1:50 000 maps. (Fig. 3.4).

This in turn indicates, among other things, that the plotting of the road network at 1:50 000 corresponds to a sampling interval of about 10 m. However, maps at this scale prove better than the simulated images in suburbs featuring narrow winding streets, although the accuracy of the images could certainly be improved by contrast and edge enhancement, something which has yet to be performed for this type of application.

The IAURIF experiments on Paris also included the merging of 20 m resolution multispectral data with 10 m resolution panchromatic data by resampling the XS1 and XS2 channels at 10 m intervals. This confirmed that the thematic content of SPOT images processed in this way is greater than that of maps at 1:50 000, and practically equivalent to that of maps at 1:25 000 as regards sizes and shapes of buildings, distribution colours and hence types of materials, etc. However, interpretation must be used with caution, particularly with regard to aberrations, wherever the object dimensions approached pixel dimensions.

An automatic classification experiment was implemented after the statistic survey entitled 'Mode d'Occupation des Sols (MOS)', or Modes of Land Use. The experiment was confined to classes having a physical representation. The following classes were identified: woods, lawns and grass-covered areas, private houses, low-apartment-type buildings, tall apartment-type buildings, freeways and major roads and streets, and, finally, water bodies and waterways. These results suggest that it may well be possible this way, in the near future, to automatically update French MOS-type surveys of urban areas.

Quick analysis of the simulated images of the two African cities reveals that 10 m resolution SPOT imagery is sufficient to gain an appreciation and to differentiate between various urban areas that cannot be distinguished with a lower ground resolution, as a result of the fact that, in certain areas of these cities, the radiometric response is scarcely different from that of the surrounding natural terrain. The poorer areas of Dakar and Ouagadougou and the major part of Agadès (Niger) are dominated by such raw natural building materials as rammed earth, thatch and straw, not to mention the unpaved roads — all of which have spectral responses that are scarcely different from each other or from that of the surrounding environment. However, with simulated 10 m resolution images, it is possible to identify the different structures and types of urban organization and hence to identify the various types of human settlements.

Finally, the inherent flexibility of SPOT data acquisition should permit the recording, even in the least favourable cases, of the one or more images per year that are necessary for monitoring the evolution of urban phenomena and for regularly updating the corresponding maps.

4.3 Coastal Studies

Coastal zones are particularly sensitive, from several points of view, and are frequently characterized by a fragile equilibrium. Thus, it is especially important that they be monitored constantly and that the development planning process take special account of a number of environmental parameters. In coastal environments, characterized as they are by continuous change and such time-varying factors as shore currents and sedimentation process, remote sensing clearly has a special role to play.

Although the SPOT system was not designed specifically for coastal studies, the first simulated images of Corsica to include a portion of coastline were soon found to be particularly rich in detail. For instance, the 10 m resolution documents clearly depicted swells aproaching the shore, while the XS1 band turned out to offer a useful degree of water penetration. Further coastal studies have since been conducted in Senegal, Bengladesh and in France (Loire estuary and North Brittany).

In each case, the results obtained have confirmed the original encouraging results. The main organizations to participate in these campaigns were : the Centre National pour l'Exploitation des Océans (C.N.E.X.O.) the Ecole Nationale Supérieure de Jeunes Filles de Montrouge (E.N.S.J.F.), the Office de la Recherche Scientifique des Territoires d'Outre-Mer (O.R.-S.T.O.M.) and the Institut Scientifique et Technique des Pêches Maritimes (I.S.T.P.M.).

One theme studied in some detail was the intertidal zone. Data analysis showed, for instance, that the intertidal zone can be extracted by a relatively simple method, that its area can be readily computed, and that distinction can be made within the zone between vegetation (algae), hard bare substrate (rock), and sand and mud. The next step was to determine the vegetation index which, in turn, yielded an accurate estimate of the total area occupied by algae. Principal component analysis of all pixels, corresponding to intertidal vegetation followed by the compilation of a colour composite of the results, made it possible to differentiate between beds of predominantly Ascophyllum Nodosum, those of predominantly Fucus Serratus and those of predominantly either Fucus Vesiculosus and Spiralis or else Pelvetta Canaliculata. The significance of this achievement is apparent when it is realized that some of these species are used in the manufacture of alginates which, in turn, are used in processed foods and cosmetics industries.

When the same analytical procedure is applied to 'non-vegetation' portions of the intertidal zone, it is possible to discriminate between bare rock (schists), sand, and other types of cover. Comparison of the results achieved by processing simulated SPOT imagery and those obtained by conventional planimetry on aerial photographs reveals differences of less than 10%. In the case of conventional false-colour photographs, it is not possible to distinguish visually between the different types of algae.

The combination of the SPOT spectral bands plus both 10 m and 20 m resolution data would thus appear to be most suitable for coastal studies. The campaign, to take place this year, covering certain coastal sites, should provide further information concerning the potential use of SPOT imagery for coastal mapping applications in such areas.

A further noteworthy point likely to be of direct interest to hydrographers is that SPOT should be capable of yielding instantaneous shorelines with an accuracy of 10–20 m.

4.4 Cartography

Standard preprocessing levels 1 and 2 might possibly prove insufficient for certain applications requiring a very high quality level of geometric correction.

Level 1 preprocessing includes only system corrections (precision of localization : 1500 m) and level 2 corrections are made using 6–8 control points per scene to present an image in a cartographic projection (precision of localization : 50 m). These preprocessing levels do not take into account displacements due to terrain relief. These distortions are negligible in the case of vertically-acquired image over flat terrain, but can be substantial in the cases of oblique views or mountainous regions.

Level 3 preprocessing uses known relief information to correct for all distortions from an existing map of the SPOT image, so that it can be overlaid on a cartographic reference model (a U.T.M. for example) to an accuracy of the order of half a pixel (10 or 5 m), whatever the acquisition angle and whatever the terrain relief.

Where maps are available, level 3 uses control points (crossroads, small lakes, etc.) clearly visible on the image, the coordinates of which are well known (within a few metres) in the three dimensions (X, Y, Z); these control points, along with orbit and attitude data, permit calculation of the image geometry. Images are then rectified pixel by pixel, based on an equidistant X, Y grid. The digital elevation model provides the Z (altitude) coordinate for each point on the grid. The radiometric value of a pixel on a corrected image is then calulated by interpolating the radiometric values of the raw image. Processing can be performed on entire image segments (up to ten consecutive scenes), thereby reducing the number of control points necessary. An orthophotograph is produced with a position accuracy of 0.5 pixel.

Use of level 3 imagery : images corrected to level 3 are of interest, on the one hand, for topographic or thematic map-making and updating, and on the other hand, for any users wishing to integrate SPOT imagery with other geographic data (geocoded data) or other space imagery (e.g. radar). Level 3 corrected images are indispensable in studying and utilizing a stereoradiometry effect of the angle of view on the spectral signature of a terrain.

Level 4 preprocessing uses two SPOT stero images to

measure terrain relief deformation and thus, to deduce and plot contour level maps.

This preprocessing level is, in a way, inversely related to level 3, for the result of level 4 is simply a representation of the relief, in the form of a digital terrain model (DTM), while that of level 3 is a corrected image.

Method: in order to determine the exact geometry of each image covering a zone to be mapped, one must calculate the precise geographic position of certain points on the ground. When the precision of existing maps is inadequate, geodetic measurements become necessary. This is effected by measuring points on the SPOT image itself, through the process of spatial triangulation, comparable to the process used in aerial photogrammetry.

Ground relief can be derived from the stereo imagery in digital (DEM) or graphic (contour levels) form. This information can be derived from photographic film using stereoplotters, or by automatic correlation procedures currently being developed.

The expected precision of altitude determination is approximately 10 m in panchromatic, for example and 20 m in multispectral.

Cartographers will be the principal users of level 4 preprocessing, as it will aid them in mapping new regions and in obtaining digital relief files more easily than by digitizing existing maps.

Nonetheless, numerous other users may find the need for digital relief data, to correct images (see level 3) and to aid in interpretation: for example, in order to correct for slope and sun exposure effects in land use studies of mountain areas.

5. CONCLUSION

The results of the simulated data studies conducted to date clearly suggest that SPOT data may well be the alternative to medium- and large-scale aerial photographs in numerous thematic applications.

The inherent acquisition flexibility of the SPOT system will be an important additional advantage in the case of environmental studies requiring multitemporal data. Moreover, the possibility of obtaining data on a cartographic projection will make it possible to go directly from data analysis to map representation. Where necessary, SPOT data will effectively permit the compilation of base maps, from the plotting of stereo pairs acquired during different passes with the imaging instrument operating in the cross-track (oblique viewing) mode. The production of maps with elevation contours accurate to within 20 m will require only 1/25th as many stereo pairs as would map production using aerial photographs at 1:100 000.

Moreover, the space imagery can be implemented using considerably simpler and cheaper equipment. Finally, the availability of image data in digital form will considerably facilitate the setting up of geocoded data banks which, in turn, will be used to generate various types of documents directly in map form.

By the time it is completed, the SPOT simulation program will have provided scientists, over a period of three years prior to the launching of SPOT 1, with data representing the full spectrum of possibilities to be offered by SPOT imagery. All of the major themes of remote sensing will have been dealt with in a wide variety of biogeographical environments, and many users will have had an opportunity of becoming acquainted with SPOT data and their potential applications.

REFERENCES

The results discussed in this paper are drawn from technical information sheets concerning SPOT simulations. These documents were prepared by scientific investigators and GDTA and are published by SPOT IMAGE.

Land Use In A Large Metropolis: Paris

Ballut, A., Laine, D. Institut d'Aménagement et d'Urbanisme de la Région Ile de France. N'Guyen, P.T. Centre Scientifique d'IBM France.

Madec, V. and Pebayle, J. Laboratoire de géographie, Ecole Normale Supérieure de Montrouge.

Torres, C.I. Groupement pour le Développement de la Télédétection Aérospatiale (GDTA).

Land Use In Corsica, A Multi Thematic Approach

Revillon, P.Y. Bureau pour le Développement de la Production Agricole.

Application of SPOT Simulated Data to the Observation of an Intertidal Zone: The Loire Estuary (France).

Belbeoch, G. and Loubersac, L. Centre National pour l'Exploitation des Océans.

Evaluation, Using Simulated Images, Of Spot Imagery Potential For Revision Of Topographic Maps

Baudoin, A., Naudin, P. and Lestringand, G. Institut Géographique National.

Spot Multispectral Data In Agricultural Statistics

Saint, G. and Podaire, A. Centre National d'Etudes Spatiales (CNES).

Fournier, P., Meyer-Roux, J. and Cordier, P. Service Central des Enquêtes et Etudes Statistiques du Ministère de l'Agriculture.

'Spot Newsletter'. No.4 issued on December 15, 1983.

4

Spacelab Metric Camera Experiments

*M. Schroeder**

ABSTRACT

Spacelab is an ideal platform for using slightly modified standard Aerial Survey Cameras in order to obtain high resolution stereoscopic photographs of the earth for mapping purposes. A first experiment of this type was carried out on the STS-9/Spacelab 1 Mission which took place from 28 November to 8 December 1983. About a thousand photographs on black-and-white and colour infrared film of various regions of the world were acquired during this mission. In total an area of about 11 Mio km^2 was covered. Due to the November launch date, illumination conditions were freqently poor over many candidate targets. However, unique high quality images with a photographic ground resolution of about 20 m were obtained. Initial image analysis has shown that these images may be used for mapping at the scale 1 : 100 000.

Because solar illumination conditions were unfavourable during the Spacelab 1 flight, NASA approved a reflight of the experiment on the Earth Observation Mission 1 (EOM-1), scheduled for September 1985. On this flight, the camera was to be equipped with a forward image motion compensation, and it was expected that a ground resolution of about 10 m would be obtained.

* Address of the author: Dr M. Schroeder, Deutsche Forschungs- und Versuchsanstalt für Luft- und Raumfahrt e.V. (DFVLR), Oberpfaffenhofen, D-8031 Wessling, Federal Republic of Germany.

1. INTRODUCTION

Since the first manned space flights photographic cameras have been used in space. The images were taken on 70 mm-film cameras, which were partly hand-held. In many cases the images were oblique photographs at scales smaller than 1 : 1 Mio. These photographs were hardly suitable for photogrammetric evalution and for mapping purposes. Their main application was therefore photo interpretation, especially in the fields of geology and hydrology.

The first images of the Earth's surface have clearly demonstrated the usefulness of space technology for observation and monitoring of our environment and have prepared the creation of systematic earth observation missions of space, especially the Landsat-programme. The geometric fidelity and the ground resolution of these space images were not precise enough for topographic mapping at scales which are in worldwide demand.

A calibrated mapping camera is an appropriate instrument for obtaining high quality images suitable for topographic mapping. The characteristics of a calibrated camera are that all distortions of the imaging geometry from the ideal central perspective are controlled and measured in the laboratory within a tolerance of ± 1 to 2 μm. These measurements can be utilized as corrections in the photogrammetric restitution of the images during the mapping process. For decades these methods have been used in aerial surveying.

The first time calibrated cameras were used in space was during the Apollo programme to map the moon's surface. However, these experiments were restricted to the moon have never been continued for the Earth. An instrument which came very close to a calibrated camera was the Earth Terrain Camera flown on the Skylab-Mission in 1973. Although this was not a calibrated camera in the strict sense, it delivered high resolution stereoscopic images which demonstrated that space photographs could be used for topographic mapping.

It is known that the U.S.S.R. has been using photographic cameras in the Sojus programme since 1976. Published photographs had been obtained with a 6-band multispectral camera (MKF-6) on small image format. However, this camera cannot be classified as a calibrated camera.

The prerequisite for the successful application of camera

systems in space is the achievement of bringing back the exposed film material to the earth. The technical difficulties that arose in the above case were at the same time also an obstacle for the continuous utilization of photographic techniques. However, because of the development of the Space Shuttle and of the growing importance of mapping of space with photographic cameras, recovery of film material from space is assured. The first step in this direction was the Metric Camera on the STS-9/Spacelab 1 mission of 28 November to 8 December 1983.

The Metric Camera Experiment on Spacelab 1 was unique since it was the first time that a calibrated mapping camera was used to photograph the earth from space. Another 'first' in space was the large image format of 23 × 23 cm.

The scientific objective of the experiment was to verify whether topographic and thematic maps at medium scale ranges (1 : 50 000 to 1 : 250 000) could be compiled from mapping camera images taken from orbital heights. Such topographic and thematic maps are required for earth resource planning and resource management on a worldwide basis. The current practice of mapping by conventional aerial photogrammetric techniques is so expensive and slow that mapping coverage of the remaining 60% of the land surface of the earth would require many years of observation. The lack of earth resource maps is particularly evident in third world regions such as Africa, South America and Asia. Mapping from space could provide an economical and efficient means by which to meet these mapping requirements. A single photograph alone covers details of several aerial map sheets.

2. THE CAMERA SYSTEM

In the field of aerial photography it is usual to use a standard film format of 23 × 23 cm. Photogrammetric evaluation equipment used worldwide is designed for this image format and for this reason was also used in this experiment.

Mainly, the application of such cameras in space depends on the correct choice of focal length. The requirements, which are imposed on the maps and the picture material taken as a basis for them, are essentially defined according to the following values:

- ground resolution (R)
- elevation accuracy (E)
- ground coverage per image (GC).

all three depend upon a given flight altitude in the following manner on the focal length (f):

$$R \sim f$$
$$E \sim 1/f$$
$$GC \sim 1/f$$

As resolution is proportional and elevation accuracy and ground coverage reciprocal proportional to f, a compromise of the focal length had to be found to fulfil the requirements for the envisaged scale range of 1 : 50 000 — 1 : 250 000. For a flying height of 250 km, the focal length of 305 mm seemed to be the best choice.

The experiment was composed of a standard ZEISS RMK A 30/23 aerial survey camera (Fig 4.1) suitably modified to interface with the Spacelab System (Schroeder 1984 a). The

Fig. 4.1 Metric camera for the first Spacelab mission.

camera was equipped with a ZEISS Topar A 305 mm f/5.6 lens with two interchangeable magazines, each loaded with 150 m of 24 cm wide film; one with a black-and-white and the other with a false-colour infrared film (CIR), (Table 4.1).

The camera was located inside the pressurized module of Spacelab. During launch and landing the camera system was stowed in experiment racks. For operation the camera was mounted in orbit over a high quality optical window by the crew. For the landing phase at the end of the mission, the crew had to remove the camera from the window and stow it again in the racks. The camera was connected to the window frame by a specially developed mount (Fig 4.2).

Table 4.1 Characteristics of Spacelab-metric camera

Type	Modified ZEISS RMK A 30/23
Lens	Topar A 1 with 7 lens elements
Calibrated focal length	305.128 mm
Max. distortion	6 μm (measured)
Resolution	39 1p/mm AWAR on Aviphot Pan 30 film
Film flattening	by blower motor incorporated in the camera body
Shutter	Aerotop rotating disc shutter (between the lens shutter)
Shutter speed	1/250 s – 1/1000 s in 31 steps
F/STOPS	5.6 – 11.0 in 31 steps
Exposure frequency	4 – 6 s and 8 – 12 s
Image format	23 × 23 cm
Film width	24 cm
Film length	150 m = 550 image frames
Dimensions: camera magazine	46 × 40 × 52 cm 32 × 23 × 47 cm
Mass: camera magazine	54.0 kg 24.5 kg (with film)

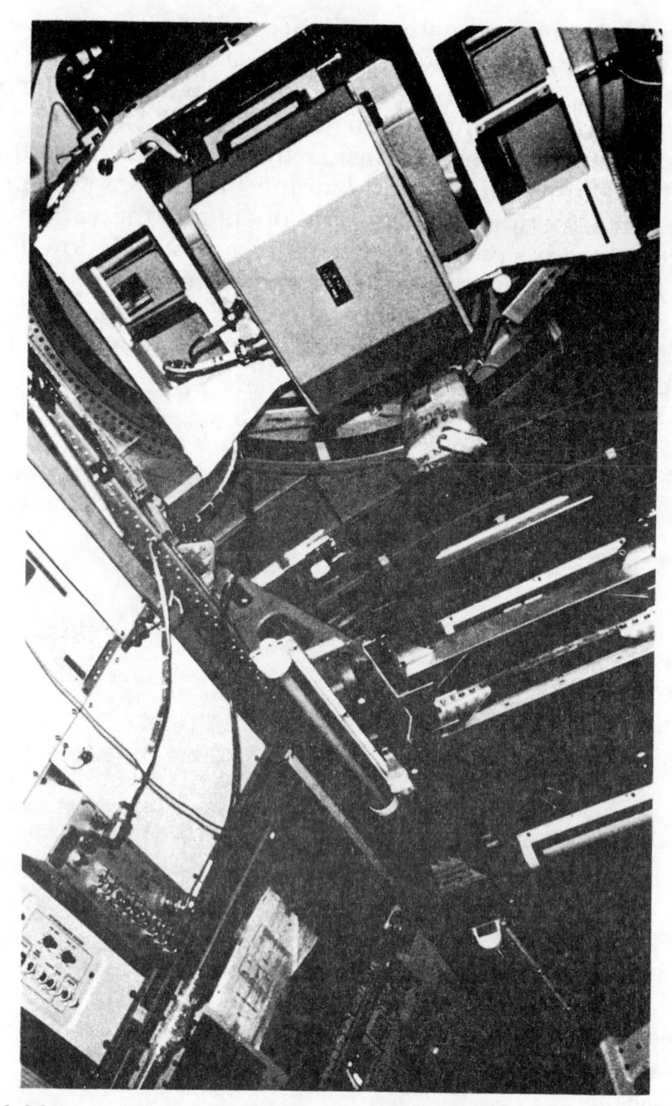

Fig. 4.2 Metric camera integrated in the Spacelab module. The camera is mounted to an optical window by a special mount.

The camera was interfaced to the onboard computer system and to the power distribution system by an electronic remote control unit (RCU). The RCU utilizes a 16-bit μ-processor to control the camera operation. The control data was loaded from the Mass Memory Unit of the onboard master computer and transferred to the camera via the RCU. The following functions could be software-commanded by the RCU or could be manually performed at the RCU control switches:

- on/off;
- start/stop of serial or single exposure;
- adjustment of 60 or 80 % overlap.

When the camera was manually operated, exposure time and aperture had to be set at the camera using adjustment control knobs. The RCU also performs:

- tagging of shutter release time;
- conditioning of housekeeping data;
- conditioning and control of power supply;
- control of auxiliary data recording.

3. EXPERIMENTAL OPERATIONS

The STS-9/Spacelab 1 System orbited around the Earth every 90 min at an altitude of 250 km. The orbit inclination was 57°, so that regions between 57°N and 57° S latitude could be photographed.

For operation of the camera the shuttle flew with the open cargo bay oriented towards the earth (Fig 4.3). The optical axis of the camera was then looking vertically down to the Earth's surface. During the 10 day mission approximately 36 h were flown in this attitude of which 4.5 h were suitable for taking photographs over land. Actually 3 h were used for camera operations to expose the whole film material that was carried in the magazine (Schroeder, 1984b).

The operation of the camera was fully automatic, i.e. start and stop of every exposure series, as well as the settings for every single exposure, were stored in the master onboard computer and were transferred to the camera via a microprocessor in the Remote Control Unit (Schroeder, 1984a).

All camera operations were ground controlled in the Payload Operations Control Center (POCC) at NASA Johnson Space Center, Houston. The instructions in the master computer could be changed within certain limits by ground commands. This was particularly important for limiting camera operations over extended cloud cover regions. As, in principle, all command updates had to be submitted for approval some hours before execution, prediction of clouds was an important part of the operation. For this an arrangement was made between ESA and the European Center for Medium Term Weather Forecasting (ECMWF) by which the ECMWF made nightly predictions of cloud cover over the target areas and transmitted them to Houston by telefax.

Fig. 4.3 Space Shuttle in Earth orientated attitude for metric camera operations. Flying height: 250 km. Ground coverage per image: 190 × 190 km.

4. DATA OUTPUT AND FIRST RESULTS

The focal length of 305 mm and the orbit altitude of 250 km yields an image scale of 1 : 820 000 which results in the 23 × 23 cm image format in a ground coverage of 190 × 190 km per image frame.

For stereoscopic evaluation the images were taken with at least 60 % overlap in flight direction. For some regions also 80 % overlapping photographs were obtained, especially over high mountain areas and over Europe. At an average ground speed of 7.5 km/s the 60 % and 80 % overlap corresponds to the shutter release event every 10 s respectively every 5 s.

Figure 4.4 shows the ground tracks where pictures were acquired during the first week of December 1983. Twenty exposure series were taken, varying in duration from 2 to 18 minutes.

Approximately 1000 photographs were taken, of which 550 were on colour infrared film and about 450 on black and white film. The colour infrared film was the Kodak 2443-Aerochrome film and the black and white film the Kodak 2405 Double — X' Aerographic film.

In total, an area of 11 million km^2 was photographed. Seventy per cent of this area was either cloud-free or only partly cloudy, and was suitable for evaluation. Images were obtained over 65 countries, most of them third world countries.

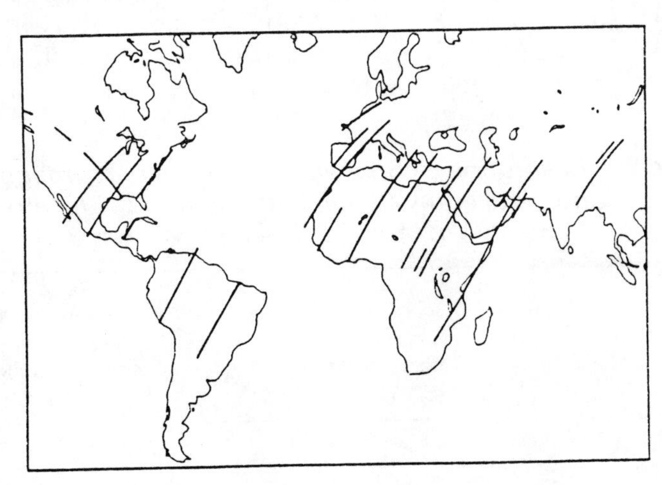

Fig. 4.4 Spacelab 1 ground tracks where metric camera images were taken.

Originally planned for the summer months, the launch had to be postponed to the end of November; thus the lighting conditions over the earth became very unfavourable as the sun elevation for all camera operations never exceeded 30°. In the northern hemisphere over Europe and North America it only reached 5 – 20°.

To compensate for these low lighting conditions longer exposure times from 1/500 to 1/250 s had to be used, while originally 1/1000 s exposure time were planned to reduce the effect of image motion during studies.

Many black and white photographs were taken at extremely low sun angles over North America. Nevertheless, they reveal the morphology of the terrain, especially in arid zones, and may be very suitable for geological studies.

The current objective of the data analysis is to investigate the scale at which topographic and thematic maps can be compiled from the images. The mapping capability of the imagery depends mainly on ground resolution, which determines the identification of objects. A first image quality control by edge gradient analysis showed that a ground resolution between 20 and 30 m was reached (Table 4.2).

First photogrammetric evaluation of the images have shown that planimetric position accuracy of less than ± 20 m can be achieved (Togliatti and Moriondo, 1984). The pointing error in height is approximately ± 20 m, which is better than the expected 0.1 % of the flying height. The images provided contour lines every 100 m for steep terrain and every 50 m for relatively flat areas. These values have proved that the metric quality of the images is suitable for mapping at a scale of

Table 4.2. Resolution of Spacelab metric camera images. For the ground pixel equivalent it was assumed: 1 lp/mm = 2.8 pixel (1 lp/mm = 1 line-pair per mm).

	Image resolution (1p/mm)		Ground resolution (m)		Ground pixel equivalent (m)	
	low contrast	high contrast	low contrast	high contrast	low contrast	high contrast
B/W-film	25	40	33	21	12	7
CIR-film	26	38	32	22	11	8

1 : 50 000. On the other hand the identification of small objects (buildings, railways etc.) which have to be plotted on a map of 1 : 50 000 is sometimes difficult due to the limited resolution and to the unfavourable lighting conditions. Due to this limitation it can be expected that the maximum map scale which can be derived from the image is 1 : 100 000, although the metric quality is better.

5. REFLIGHT ON EOM-1

Because the solar illumination conditions were unfavourable and the number of observations were limited during the Spacelab 1 flight, NASA has approved a re-flight of the experiment on the Earth Observation Mission 1, which was scheduled for September 1985. In this mission a Spacelab was again to be flown as an experiment platform. The same interfaces between the camera and the Spacelab system would be used as in the previous mission, and even the orbit parameters such as flight altitude and inclination would be the same. The mission duration would be 7 days.

The main differences with Spacelab 1 are:

- 3 film magazines with a total amount of 2000 images will be used to improve the global coverage;
- the magazines will be equipped with a forward (image) motion compensation (FMC) to improve the resolution.

For cameras with FMC, as in Spacelab 1, the resolution is limited by image motion in the focal plan, which is due to the relative volocity between the camera and the ground.

For a flying height of 250 km and a focal length of 305 mm the image moves with a velocity of 9 mm/s in the focal plane. In Spacelab 1 short exposure times of about 1/500 s were used to minimize this motion. But for short exposure times relative sensitive films of medium to low resolving capability have to be used. This limited the resolving power of the Spacelab 1-camera system (lens, film motion) to approximately 25–40 lp/mm (Table 4.2).

In the EOM-1 mission the image motion was compensated for by moving the film in flight direction with the same velocity as the image motion. This allows longer exposure times and the use of slow, but high resolution films. By this means a resolution of 60–80 lp/mm corresponding to 10–13 m at the ground can be expected.

6. CONCLUSION

For all the countries of the earth the availability of up-to-date topographic maps is an essential prerequisite for the use and further development of their infrastructure and for optimum management of their natural resources. But only in a few countries are such maps available. The conventional methods used so far in mapping are not capable of covering the worldwide demand. Metric camera images from Spacelab 1 have demonstrated that topographic mapping at a scale of 1:100000 is possible from space images. In spite of the unfavourable lighting conditions during the Spacelab 1 mission, it was possible to produce the best images of the Earth's surface that have been obtained from space for photogrammetric evaluation.

In the NASA-EOM-1 mission an improved camera system of the same type as in Spacelab 1 was flown. This camera was equipped with a forward image motion compensation to produce images of approximately 10 m ground resolution. It is expected that these images are suitable for deriving maps at a scale of 1 : 50 000.

Up until now the application of calibrated photogrammetric cameras in space was an area which had been neglected. The goal of the two Metric Camera Experiments is to demonstrate that this gap can be filled by utilizing the extended capabilities of the Space Transportation System and that the obtained images can effectively contribute to an improved cartographic coverage of the earth.

References

Schroeder, M. 1984a. The Metric Camera Experiment on the first Spacelab Flight. *Acta Astronautica*, (1) 73-80.

Schroeder, M. 1984a. Flight performance of the Spacelab Metric Camera Experiment. *Proc. IGARS' 84 Symp.* Strasbourg 27-30 August 1984, Ref. ESA SP-215.

Togliatti, G. and Moriondo, A. 1984. Analysis of the metric camera B & W images over Italy. *Proc. IGARS' 84 Symp.* Strasbourg 27-30 August 1984, Ref. ES SP-215.

5

Coastal Zone Color Scanner (CZCS) and Related Technologies

*Buzz Sellman**

ABSTRACT

The Coastal Zone Color Scanner (CZCS) is an experimental programme of the National Aeronautics and Space Administration (NASA), and was the first space-based mission devoted to the measurement of ocean color and temperature. As its name implies, the mission was designed to collect data over the world's coastlines, as these waters are a significant source of planktonic plants that are crucial to the total ocean food chain. Since its launch in 1978, numerous results have verified that the experimental goals were achieved, and the stage has now been set to initiate global oceanic observations that can build upon the findings and insights gained from the CZCS mission. This chapter briefly reviews the CZCS programme and offers the participants at this meeting some suggestions on how they can improve upon the benefits they might receive in their respective countries as a result of research programmes like CZCS.

* Address of the author: Dr Buzz Sellman, Environmental Research Institute of Michigan, Ann Arbor, MI 48107, U.S.A.

The CZCS was launched aboard the NIMBUS-7 spacecraft in October 1978. The NIMBUS-7 satellite contains eight individual sensor packages, and each is dedicated to a specific experimental mission. The eight sensors/experiments are as follows:

- Coastal Zone Color Scanner (CZCS);
- Scanning Multichannel Microwave Radiometer (SAMS);
- Earth Radiation Budget (ERB);
- Stratospheric and Mesospheric Sounder (SAMS);
- Stratospheric Aerosol Measurement II (SAMS II);
- Limb Infrared Monitor of the Stratosphere (LIMS);
- Solar Backscatter Ultraviolet/Total Ozone Mapping Spectrometer (SBUV/TOMS);
- Temperature Humidity Infrared Radiometer (THIR).

The CZCS was the first NASA space sensor dedicated solely to studies of marine resources, specifically visible measurements of ocean color as it relates to primary productivity and surface temperature measurements. Starting in 1976 NASA convened a Nimbus Experiment Team (NET) for CZCS, and this group was charged with planning the overall scientific research programme for CZCS. With the successful launch of NIMBUS-7 in late 1978, an analysis programme was begun by researchers (primarily from the United States, Europe and Japan) working on test sites over the entire globe.

The NIMBUS-7 spacecraft characteristics and the CZCS spectral bands are summarized in Table 5.1.

Table 5.1. Selected Nimbus spacecraft and CZCS characteristics

Orbit altitude	— 955 km
Equator crossing	— 12:00
Nadir ground resolution	— 825 m
Swath width	— 1566 km
Field of view	— ± 39°
Spectral Bands	
1 — 0.43 — 0.45 μm	— blue
2 — 0.51 — 0.53 μm	— blue/green
3 — 0.54 — 0.56 μm	— green
4 — 0.66 — 0.68 μm	— red
5 — 0.70 — 0.80 μm	— near i.r.
6 — 10.5 — 12.5 μm	— thermal i.r.

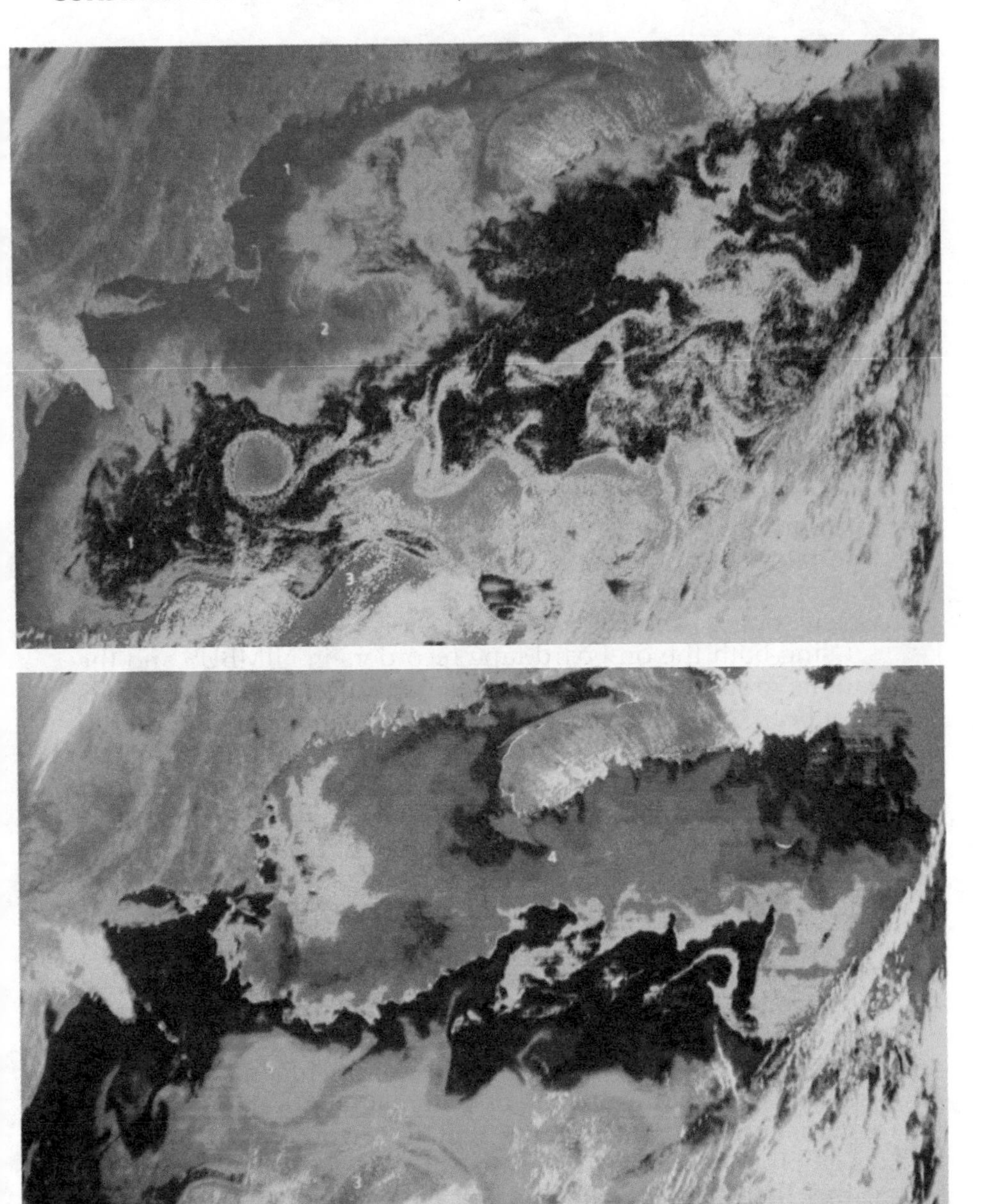

Fig. 5.1. Processed CZCS data of chlorophyll concentrations (top) and surface temperatures (bottom) along the U.S. Atlantic coast.

The placement of the CZCS bands was designed to measure varying concentrations of phytoplankton pigments in the ocean. For high concentrations of pigments (chlorophyll a), the measured signal in the blue (0.44 μm) and red (0.67 μm) bands will decrease, due to the absorption properties of these pigments. At the same time, the measured backscatter in the blue/green (0.52 μm) and green (0.55 μm) bands will increase as pigment concentrations increase. Thus, water that is essentially free of phytoplankton pigments will appear a deep blue, while waters with high concentrations of these pigments will appear green.

Figure 5.1 demonstrates very well these band sensitivity differences. In the upper image, the chlorophyll-rich shore zone of the Eastern coast of the United States (reddish tones) is very distinct from the blue Gulf Stream waters, which are poor in chlorophyll. The lower images shows the surface temperature patterns at the same time. The front between the colder waters to the north (blue tones) and the warmer Gulf Stream waters (orange and yellow tones) is quite dramatic.

Using both the on-board tape recorders in NIMBUS and the various ground receiving antennae to collect data, researchers have been analysing coastal and open ocean waters in order to begin to understand the dynamics of phytoplankton concentrations on a global basis. This work could eventually offer direct benefits to developing countries, but at the moment it is difficult to show or when that will happen. The reasons for this are varied.

Perhaps the most obvious point to return to is the fact that the CZCS programme is a research programme of NASA. It began in the early 1970's and the current phase of this programme, which it is finished, will have involved perhaps 15 years of research. This period has been used to build knowledge, establish and improve experimental methods, and occasionally show examples of potential applications through demonstrations. The benefits that this work might have for developing countries' needs are not central to the programme at this time, and one should not expect that they would be. Neither NASA nor NOAA have a specific charter to incorporate developing country needs into their R&D plans and progammes, although occasionally they do promote international cooperation on these programmes if developing countries are prepared to finance that participation. Yet in order to increase the value of programmes like CZCS to developing countries, one needs, among other things, information, patience and money.

The patience and money required to realize useable benefits from programmes like the CZCS is simply a realization that space-based research programmes that attempt to understand global cycles and systems are extremely complicated. The CZCS program to investigate and better understand primary productivity in the world's oceans is a good example of a worthy, but complicated, research enterprise.

REFERENCES

Browder, J.A. and Powers, J.E. (eds) 1980. *Proc. Coastal Zone Color Scanner Workshop*. NOAA. Mem. NMFS-SEFC-9, Miami, Florida, February 1980.

Brown, O.B. and Cheney, R.E. *Adv. Satell. Oceanogr. Rev. Geophys. Space* Phys. 21,5.

Caraux, D. and Austin, R.W. 1983. Delineation of seasonal changes of chlorophyll frontal boundaries in mediterranean coastal waters with NIMBUS-7 coastal zone color scanner data. *Remote Sensing Environment* 13, 239—249.

Colwell, J.E., 1983. Regional inventory by joint use of coarse and fine resolution satellite data. *Proc. 17th Intl Sym Remote Sensing Envir,* Ann Arbor, Michigan, May 9-13, 1983.

Esaias, W.E. 1981. Estimating Oceanic Primary Productivity from Space, JPL, California Institute of Technology.

Fusco, L. NIMBUS-7 coastal zone colour scanner data processing for Earthnet-experience to date. *Eur. Space Agency* Bull, 27.

Gordon, H.R. *et al.* 1983. Phytoplankton pigment concentrations in the middle Atlantic bight: comparison of ship determinations and CZCS estimates. *Applied Optics*, 22, (1).

Gower, J.F.R. (ed.) 1980. Oceanography from space. *Pro. COSPAR/SCOR/ IUCRM Sympo*. Venice, Italy, May.

Hovis, W.A. *et.al.* 1980. Nimbus-7 coastal zone color scanner: system description and inital imagery. *Science*, 210, 3 October.
J. Geophys. Res. 89, June 1984.

Robinson, I.S. 1983. Satellite observations of ocean colour. *Phil Trans. Royal Soc. Lon*, A 309, 1508.

Tanis, F.J. and Lyzenga, D.R. 1981. Development of great lakes algorithms for the Nimbus-G coastal zone color scanner. *ERIM* Report No. 150000-11-F, June.

Tanis, F.J. and Knorr K.H. 1980. Use of airborne data to suppose validation of the coastal zone color scanner in the Gulf of Mexico. *ERIM* Report No. 138700-1-F, December.

Walsh, J.J. (ed.) 1982. The marine resources Experiment Program (MAREX). Report of the Ocean Color Science Working Group, NASA/GSFC, December.

6

Selected Features of the SEASAT Satellite

*R. Keith Raney**

ABSTRACT

This paper is meant to describe selected features of the SEASAT Satellite and gives further information on the sweep of imaging radar capability worldwide.

1. BACKGROUND

The principal purpose of this section is briefly to describe the milieu of imaging radars. The family of relevant radar systems is outlined spanning the years from 1978 to 1990.

Prior to SEASAT, there were generally two Real Aperture Radar (RAR) side-looking radar systems available, the Westinghouse APQ-97 and the Motorola APS-94. Some X-band SAR data from Goodyear radars was available prior to 1978, as well as experimental data at X- and L-band from the University of Michigan Willow-Run Laboratories (which became ERIM in 1973), and from the Jet Propulsion Laboratory at L-band.

SEASAT remains a major milestone in imaging radars. After SEASAT in 1978, the next free-flying satellite SAR systems will be the European Remote Sensing (ERS-1) Satellite, the Earth Resource Satellite of Japan (J-ERS-1) and Canada's RADARSAT, all appearing more than a decade after SEASAT.

Shuttle radars SIR-A, SIR-B, SIR-C provide exciting ex-

*Address of the author: Dr R. Keith Raney, Chief Radar Scientist, Canada Centre for Remote Sensing, 2464 Sheffield Road, Ottawa, Ontario K1A OY7, Canada.

perimental interim capability, but remain very limited in time and space. (A comparison of civilian imaging radars is given in Table 6.1.)

Table 6.1 Imaging radars (civilian).

Aircraft RAR		
APS-97	(K)	
APS-94	(X)	(1975–)
Aircraft SAR		
JPL	(L)	(1970–)
SAR-580	(L,C,X)	(1978–1984)
Aero Service	(X)	(1975–)
STAR-1	(X)	(1983–)
C-IRIS	(C)	(1985–)
Aircraft (R/D)		
China	(X)	(1981–)
Vigie (France)	(X)	(1975–)
Denmark	(X)	
Russia	(X)	(1983–)
Shuttle		
SIR A	(L)	(12–14 NOV 1981)
SIR B	(L)	(OCT 1984)
SIR C	(L,C,X)	(1989 ?)
Earth Observation		
SEASAT	(L)	(JULY–OCT 1978)
ERS-1 (ESA)	(C)	(1989–1991)
J-ERS-1 (Japan)	(C)	(1989–1991)
RADARSAT (Canada)	(C)	(1990–1995)
COSMOS (Russia)	(X)	(1983–)
Extra-terrestrial		
Apollo 17	(VHF)	(1974)
Pioneer Venus	(S)	(1978–82)
Venera (Russia)	(X)	(1983)
Venus Radar Mapper	(S)	(1988–89)

2. VIEWPOINT

Imaging radar has a superficial similarity to aerial photography. Interpretation of radar imagery is aided by the trick suggested in Fig.6.1. The illumination is oblique, coming as it does from the side of the scene. The appropriate viewing direction is from above the image, at right angles to the transmitted microwave energy. Prespective from this viewpoint is created by the progressive time delay proportional to distance in the scene from the radar. Indeed, first order

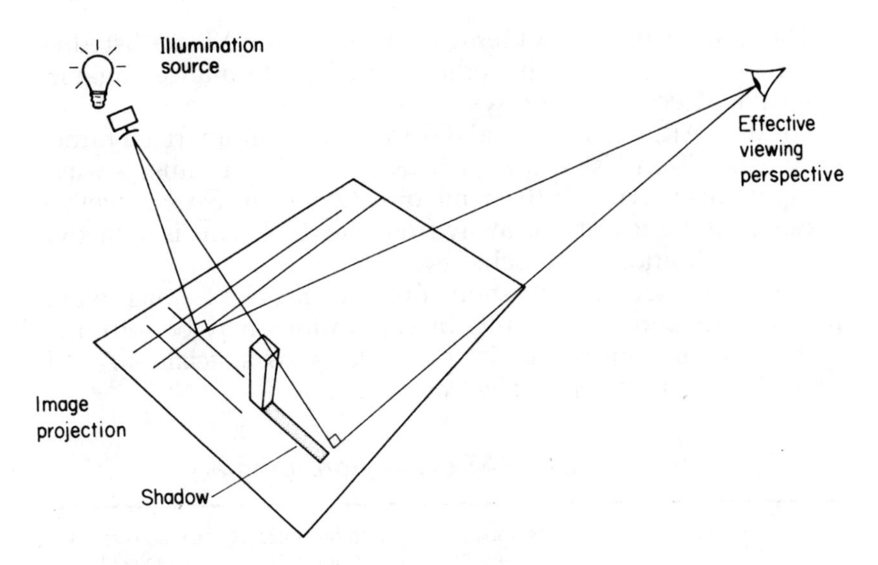

Fig. 6.1. Interpretation aid for radar imagery.

geometric effects such as elevation displacement become equivalent between radar and aerial photography in this frame of reference, as reflections from taller objects are shifted away from the viewpoint nadir, thus towards the radar in the side-looking geometry.

3. SEASAT

The SEASAT satellite system became operational early in July 1978, and ceased operation (due to a short circuit in the solar panel slip ring assembly) on 10 October 1978. The NASA satellite was the result of special expert panels and a Johns Hopkins University study during the early 1970's, and was built through a JPL Project Team. Its importance to the development of civilian space radar programmes and enthusiasm for microwave remote sensing cannot be underestimated.

Unfortunately, the spatial coverage from the SEASAT SAR was limited by direct transmission of telemetry data to four North American receiving stations and one station in England. Areas outside these regions received no SEASAT SAR coverage.

The microwave instruments on board SEASAT are listed in Table 6.2, together with demonstrated performances against principal objectives of the system.

For the SAR, wave spectral information is summarized rather optimistically in Table 6.2. A SAR is not able to image wave components waving in the same direction as the SAR as well as waves moving toward or away from the SAR. This is an active area of applications research.

More detailed information on the SEASAT microwave instruments and their results in applications may be found in many sources, including in particular two special SEASAT issues of the *Journal of Geophysical Research*.

Table 6.2. SEASAT evaluation summary.

Sensor	Observable	Demonstrated Accuracy (1δ)	Demonstrated range of observable
Altimeter	altitude	8 cm (precision)	$H_{1/3}$ <5 m
	significant wave height ($H_{1/3}$)	10% or 0.5 m	0 to 10 m
	wind speed	2 m/s	0 to 10 m/s
Scatterometer	wind speed	1.3 m/s	4 to 26 m/s
	wind direction	16°	0 to 360°
Scanning multichannel temperature microwave radiometer	sea surface temperature	1.0°C	10 to 30°
	wind speed	2 m/s	0 to 25 m/s
	atmospheric water vapour	10% or 0.2 g/cm^2	0 to 6 g/cm^2
Synthetic aperture radar	wave length	12%	wave length ⩾100 m
	wave direction	15°	0 to 360°

Note: demonstrated accuracy, as used here, refers to SEASAT data compared with conventional or *in situ* measurements within the stated value.

4. SAR TECHNOLOGY

The most unique instrument on SEASAT was the SAR. In this section we outline the principal features of SAR systems at an introductory level.

The objective of the discussion is to show that SAR systems, when well designed and implemented with modern digital technology, are not especially complicated devices from the user's point of view.

An imaging radar system actively observes the scene of interest by measuring back scattered microwave energy. The principal parameter for distributed reflecting objects is sigma nought, a measure of average local reflectivity. The output is a map of various brightnesses, whose variations contain the information of interest. These variations may in many cases be reduced through further processing to geophysical parameters of interest.

Imaging radars, unlike vertical aerial photography, rely on different mechanisms to form the image in each of the two dimensions. Range information is gathered at essentially the speed of light, whereas the along-track (azimuth) information is gathered at the speed of traverse of the sensor, like any other scanning system. SEASAT will be used as an example for these discussions.

There are three logical steps involved in the practical operation of an SAR. The radar itself, on a spacecraft or an aircraft, gathers signals from all of the terrain reflecting energy back to the system. The processor, usually in a ground-based facility, integrates over the data. Digital processors operate in a fashion analogous to their optical predecessors; they focus the signals to form images of the terrain reflectivity map. The image data is usually subjected to further analysis, either quantitative or qualitative, to derive information (Fig.6.2.).

For a good system, the radar and the processor should be invisible to the user, imposing only second-order artefacts onto the image. SEASAT (when digitally processed) is a rather good system. The STAR-1 (Intera — MDA — ERIM) is a very good system, in the sense used here.

The radar determines in a fundamental way the basic image quality, such as resolution, swath width and noise characteristics.

The processor is the primary limitation to timely availability of the image.

Discriminating users always want more data of higher quality (Fig.6.3). For an imaging radar, this translates into better resolution, wider swath widths, and more averaging to reduce the multiplicative noise peculiar to coherent (SAR) imaging systems. The term used in SAR literature to express the effective number of independent averages employed is

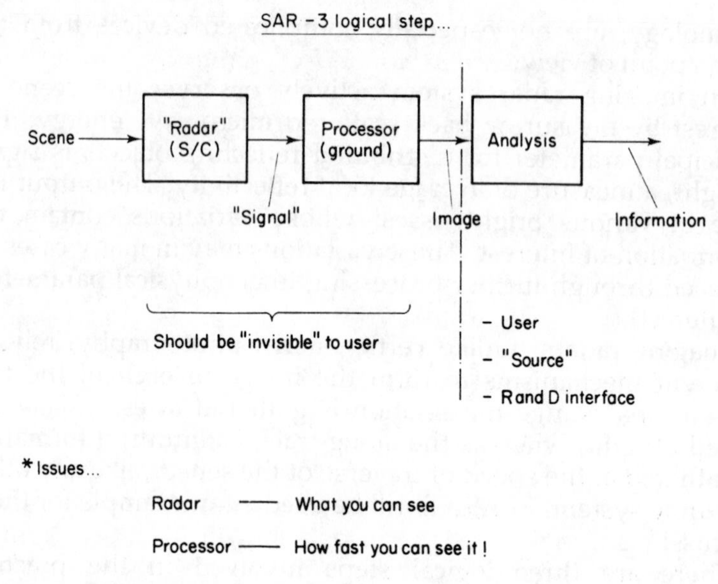

Fig. 6.2. Logical steps involved in the practical operation of an SAR.

	Swath (Slant Rate)	ΔR_s
	Resolution (RXA)	$r\ (r_{gr} \times r_a)$
	Looks (Speckle Reduction)	N
→	Ambiguity Restraint	
	$\frac{N \Delta R_s}{r} \leq \frac{c}{4V}$	(Practical Limit)
→	Power Restraint:	

$$N^2 \Delta R^2 / r_a^2\, r_{gr} \leq \overline{GF}$$

GF → User 'Greed Factor'
Set by Available Power from Radar Hardware
c = Speed of Light
V = Speed of SAR Vehicle

Fig. 6.3. User requirements and its technical implications.

'looks'. Basic SAR system operation imposes fundamental upper bounds preventing unlimited simultaneous improvements of these factors.

In order to avoid the appearance of mutiple exposures (ambiguities) in SAR imagery, the radar sampling frequency must be both large enough *and* small enough. Ambiguities threaten to dominate image quality if the restraint is not satisfied. (For SEASAT the inequality is just satisfied.)

In order to avoid decreases in the average reflected signal relative to the background noise (which is always present in active imaging systems), the image quality parameters must also satisy an upper bound as set by the available power to be transmitted by the radar. Rearrangement of resolution, swath width and number looks is possible (at the time of design) within these restraints.

The task of the processor is to integrate over the signal corresponding to one reflecting element in order to form its image, and to do this for all resolution cells in the scene. Each reflector returns a signal while within the instantaneous field of view (IFOV) of the instrument, which for a radar is given by the product of the pulse length (in range) by the width of the antenna pattern (in azimuth). For SEASAT, the IFOV is 13 × 18 km. The resolution cell (for each of the four looks) is 25 × 25m. The ratio of these areas is approximately 400 000 — a large data processing task for the processor.

The task is large . . . How fast can it be done? For SEASAT there has been a marked increase in processing speed, as illustrated in Fig.6.4. Operational spacecraft systems of the future will have processing systems that will produce images much faster than the related infrastructure can handle. Fast processing speeds have been demonstrated in the laboratory. For the STAR-1 airborne SAR, image formation is essentially real time, and built into the system.

In spite of the subtleties of radar operation and the magnitude of data processing manipulations, the result of these two sub-systems is remarkably stable and predictable (over the linear operating range of the system elements). This stability may be represented by two fundamental laws of conservation, as expressed in Table 6.3.

Now we have a SAR system reduced to its fundamentals. The output of a SAR in response to varying terrain features is a map of different brightnesses over the image. What features of

	(ONE SCENE)	
1978	40 Hours	(MDA)
1984	2 Hours	(JPL)
1990	1 Minute	(ERS-1, RADARSAT)

*CONCLUSION?

USER SHOULD VIEW SAR DATA AS (NEAR) REAL TIME IN 1990'S.

ALSO

PROCESSING SPEED (AIRBORNE SAR)
MODERN DIGITAL SYSTEM —
eg. *STAR-1* ⇨ *REAL TIME!*
(INTERA — SYSTEM OPERATION
MDA — ON BOARD PROCESSOR)

Fig. 6.4. Processing speed of SAR data.

Table.6.3. Conservation principles.

Processor and analysis

Conservation of energy
Mean reflectivity (σ^o) is independent of number of looks, time variations in scene reflectivity, or processor 'focus'

Conservation of confusion
Resolution is degraded proportional to decrease in speckle variance (in the processor, or in analysis)

Interpretation
Spatial variations in reflectivity ('information') are not compromised by thoughtful processor and analysis operations

the scene, or the radar and scene in combination, result in different brightness?

Table 6.4 lists (in order of importance in each group) the most important determinants of reflectivity. Surface roughness (relative to the radar wavelength, as projected onto the surface) is the most important parameter. Since microwave systems use wavelengths on the order of size comparable to geophysical features of interest (foliage, alluvial deposits, capillary waves, etc.), the brightness of reflected energy often may be related to information of direct benefit to the user.

Penetration of surface media occurs only for very favourable values of dielectric constant, such as occurs for extremely dry soils observed at long wavelengths, or for non-dense media such as certain agricultural crops.

In general, all of these parameters are interdependent ... which makes life interesting for microwave applications scientists.

Table 6.4 Image brightness ~ reflectivity[a]

(i) *Reflectivity* FNC *of*
   [a]*Roughness* WRT λ (wavelength)
   Dielectric constant
   Motion

(ii) *Radar/Scene Geometry*
   [a] *Incidence Angle* (WRT vertical)
   Aspect Angle
   Wave length
   Polarization
   Terrain

[a]For natural or distributed scenes such as forests, agriculture, or oceans.

5. EXAMPLES

SEASAT SAR examples are well-known, and may be found in JPL publications, or several recent books, as well as selected reports and professional papers.

Digitally processed SEASAT SAR products have shown very good geometric fidelity, particularly the MDA results. Many examples have been created of SEASAT SAR data merged with Landsat or other sensor data. (Examples: JPL, Goodyear, CCRS, DFVLR, etc.)

SEASAT was not a calibrated SAR, though it has contributed to an appreciation for and understanding of the problems involved. Users should insist on calibrated systems in the future.

The scatterometer on SEASAT was used to measure surface wind speed and direction over much of the globe by microwave measurements of the reflectivity of the ocean surface. On average, the reflectivity is proportional to roughness at the scale of the probing wavelength which was for this instrument approximately one centimetre. Since such ocean roughness is very responsive to local wind conditions, wind speed may be estimated.

Wind direction may be inferred by measuring the apparent reflectivity in two or more directions, and by solving an empirical relationship. Unfortunately, this leads to several possible wind directions, the famous directional ambiguity problem for this instrument.

Fig.6.5 shows the geometry of radar beams used by SASS to probe the ocean directional reflectivity. Scatterometers used on other satellite missions have related antenna pattern distributions.

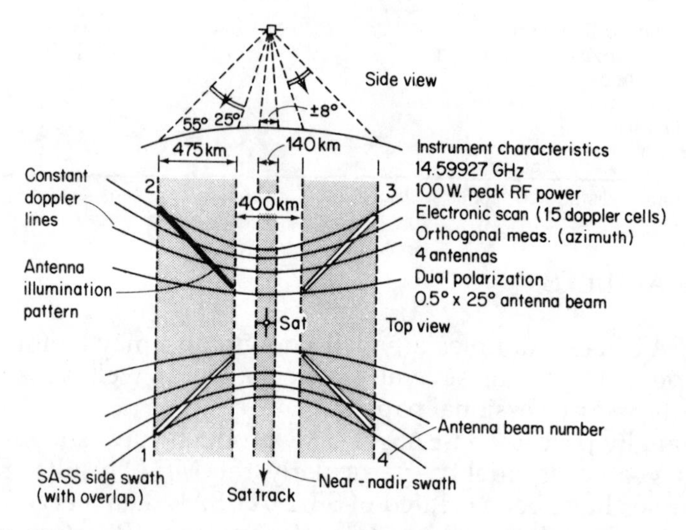

Fig. 6.5. SEASAT scatterometer illumination patterns and instrument characteristics.

The results of SASS measurements and independently observed wind speeds for the JASIN experiment are outlined in Fig.6.6. The comparison is quite satisfying.

Part of the scatter in data points can be explained by noting that the two kinds of measurements are different. SASS measures a spatial average of wind speed as it directly impacts the sea surface, whereas the surface observations are single point time averaged measures, nominally (but never actually!) at 10 m elevation above the sea.

Typical variations in possible solutions to the wind direction algorithm are shown in Fig. 6.7. Each solution has a probability associated with it. Unfortunately, often the most probable solution is not the correct solution. A solution to the ambiguity problem seems to have been developed, as seen in the next discussion.

By using a method of digital spatial vector consistency analysis, the British Meterological Office (Bracknell; Dr D. Offiler) has developed a solution to the ambiguity problem. The process requires only a few minutes of machine time, and has shown nearly perfect vector selection for the test cases

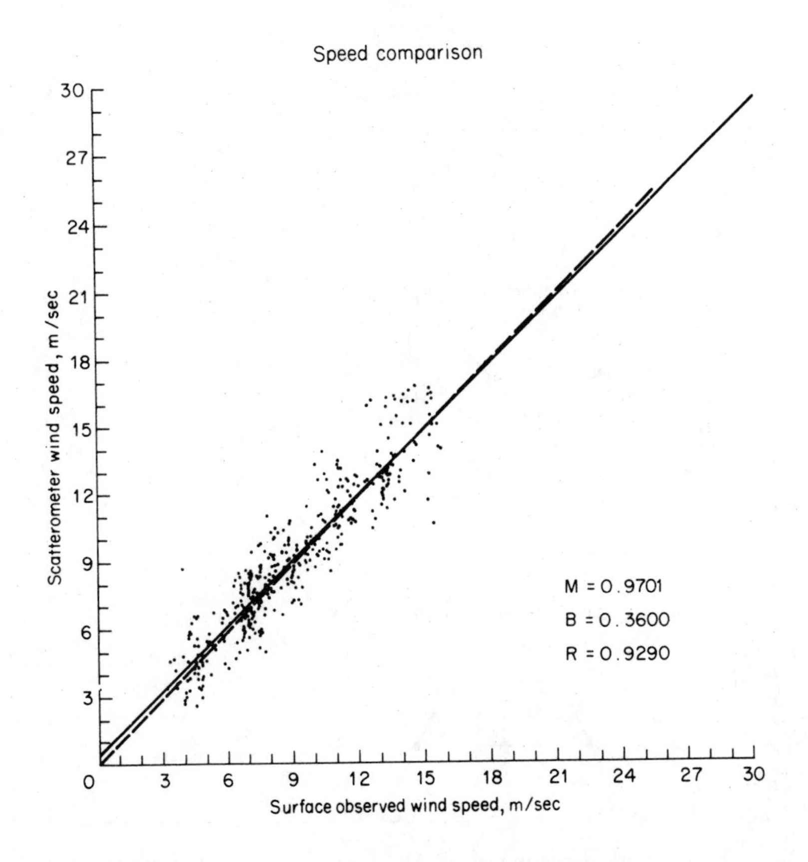

Fig. 6.6. Comparisons of SASS wind speed solutions with surface wind speed observations in the JASIN experiment.

examined. The method seems to have the performance and speed required of an operational system. The name Profiler has been proposed for this algorithm.

Because of severe visibility problems and large coverage requirements, Canada has embarked on a programme to achieve a free-flying remote sensing satellite. The system is meant to be operational (following a qualification phase) so that efficient and rapid data processing is required. A tape recorder will be carried, capable of storing SAR data for later replay to an

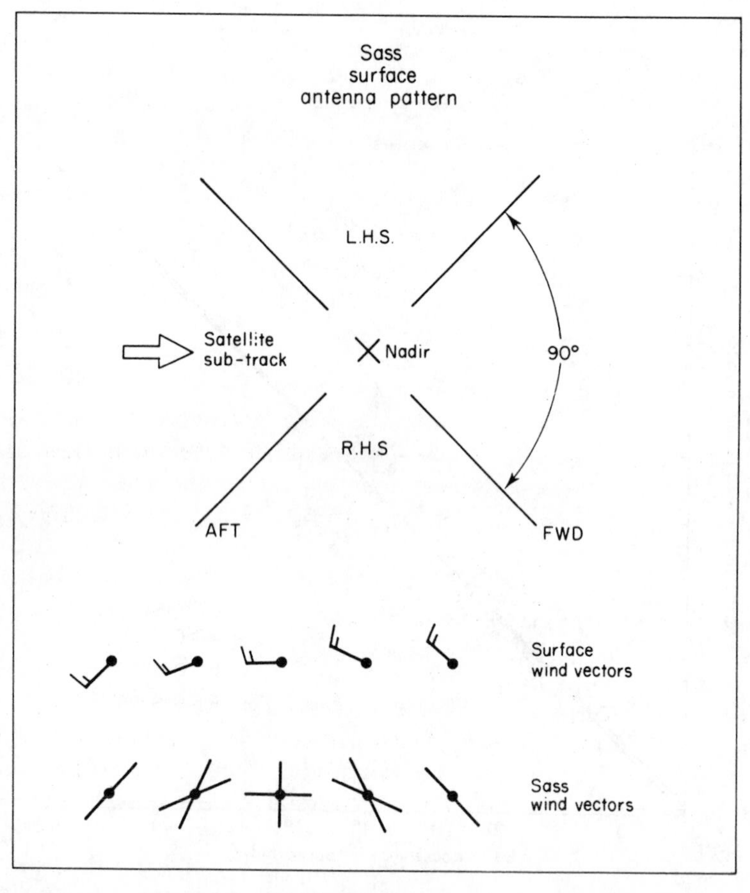

Fig. 6.7. Schematic depicting the SEASAT antenna illumination pattern on the earth along with five possible wind directions and the corresponding solutions produced by the SASS wind algorithm.

Earth receiving station, thus allowing potential for complete global SAR coverage.

The system will be operated on a basis similar to that employed by NOAA (ex-NASA) for the Landsat series of remote sensing satellites, with ground-station fees and data-product costs to be borne by the users. Ground stations will be encouraged. In addition to national objectives, including ship navigation in the Arctic, global objectives include geological structures and tropical forests. It is anticipated that an international announcement of opportunity will be released toward the end of the decade for scientific participation in the project.

The total Radarsat concept includes the satellite itself, X-band communication links to the receiving stations, thence (for Canada) to the mission control and applications information centre, rapid data processing, combination with ancillary data to create a map-like information product which is then relayed to high-demand users.

Applications with less dynamic requirements will be served with less rapid data product generation. Enhancement and specialized analysis will be the responsibility of various user centres.

Scatterometer data will regularly be processed and released through Environment Canada.

Unlike other satellite SAR's that look to the north of the sub-satellite track, Radarsat is being developed to look to the south. This provides coverage of the active Arctic area several times a day, otherwise an impossibility with only one satellite. Current design for Radarsat suggests that a maximum swath width of 120–140 km will be achieved. The SAR on Radarsat will look south throughout the orbit, thus providing nearly complete coverage of Antarctica, probably the first imaging radar to do so.

For areas of the world already covered by SEASAT, ERS-1, or the shuttle missions, the viewing geometry of Radarsat will provide an additional perspective. Radar is designed to cover the range of incidence angles from 20° (near edge) to 45° (far edge), by electronically charging the elevation angle of the antenna beam at the radar. Thus four positions, each of a ground swath width of approximately 130 km, will be sufficient to access anywhere within a 500 km wide domain during any satellite pass. Incidence angle may be chosen to optimize response for a particular application. Stereo SAR imagery may be developed for eological or cartographic purposes, and site revisit time may be reduced.

The Radarsat schedule is shown in Fig.6.8. Several key milestones have been passed, the most recent being approval and initiation of Phase B, the detailed design and costing of the system, together with hardware development. The next critical event is submission of the Phase C/D proposal by the project to the Cabinet, scheduled for the autumn of 1985. (If all goes according to plan, approval is anticipated after a one-year consideration period: thus the schedule shows an autumn 1986 start for the construction/implementation phase.)

The Radarsat project would benefit from international expressions of support, and enquiries regarding ground station participation and data access.

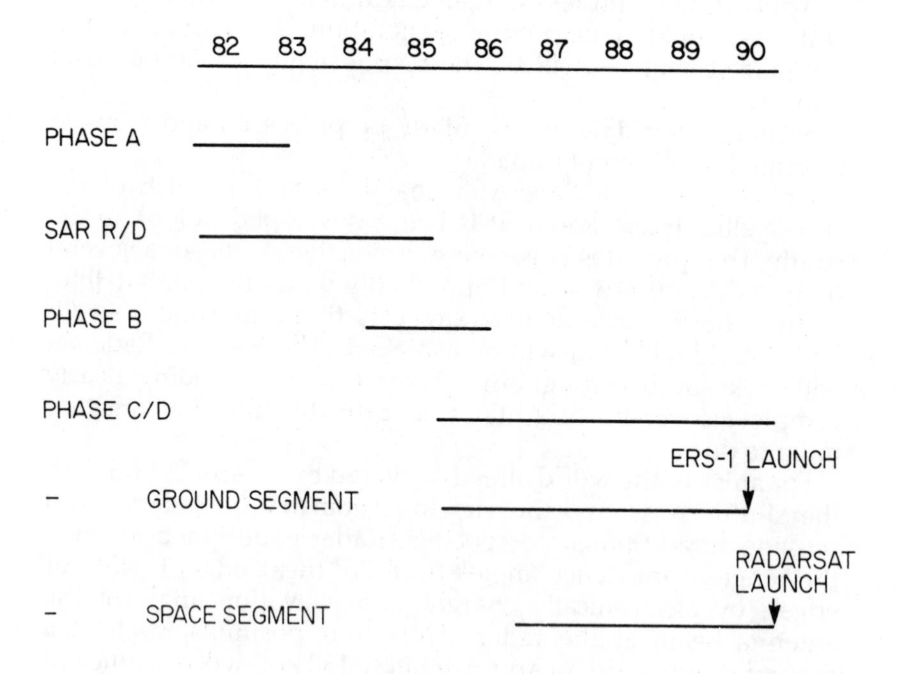

Fig. 6.8. Radarsat schedule.

6. ACKNOWLEDGEMENTS

The remarks in this paper span the work of many individuals and organizations. The author is indebted to JPL, DFVLR, NASA, Marconi (UK), MDA (Canada), the British Meterorological Office, ESA (Europe), CCRS (Canada), and the Radarsat Project for many contributions and constructive dialogues with colleagues, and for permission to use materials in this and related presentations.

7

First Results of the European Spacelab Photogrammetric Camera Mission

*Gottfried Konecny**

Photogrammetry (photography from aircraft) is now considered the standard topographic data acquisition tool the world over and yet, progress in mapping from aerial photographs is so slow that it has accomplished no more than 40 — 45% of the desired topographic map coverage at the necessary scales of 1:50 000 or 1:100 000. With existing methods of aerial photogrammetry, it will not be possible to achieve world land coverage at these scales by the year 2000.

In 1972, the first Landsat introduced satellite imagery to most countries of the world. However, Landsat film products, available from the U.S.A. and from various receiving stations all over the world, have not been able to meet the resolution requirements to provide sufficient detail for topographic cartography.

Even the more costly process of using digital computer tapes with computer enhancement techniques was not able to deduce objects smaller than 200 m from the 79 m pixel data. Transportation networks, essential to topographic mapping, could therefore not be detected in the images.

As of 1982 with Landsat 4 and 5, we have the possibility of using 30 m pixels with thematic mappers. Evaluation of these

* Address of the author: Professor Gottfried Konecny, University of Hannover, Institute for Photogrammetry and Engineering Surveys, Nienburger Str. 1, D-3000 Hannover, Federal Republic of Germany.

images requires a computer effort about ten times as high as for Landsat MSS, known from earlier missions. The object detection capability limit of the thematic mapper images of 30 m pixels is about 80 m on the ground.

Already in 1973, SKYLAB photography had rendered better resolution with an image capability of 40 m or a pixel equivalent of 15 m.

Given the requirements for topographic mapping at the scale of 1:50 000, we have tried at the University of Hannover to determine the requirements for digital and photographic sensor resolution.

The enlargement of a digital Landsat MSS image, compared to a 1:50 000 topographic map, shows clearly, that Landsat MSS is not suitable topographic mapping at that scale. However, 1:50 000 aerial photography can, as we know, be used for mapping at that scale.

We have digitized aerial photography at the scale of 1:50 000, with an Optronics drum scanner at ground pixels of 1m, 2.5m, 5m, 10m, 20m and 40m etc. not only for single images, but also for overlapping stereo-images. We tried to evaluate these stereo-pairs, digitized at various intervals, and found that, for European topographic mapping standards, a pixel size of 2.5m is required in monoscopic observation to discern roads, buildings, hydrology and vegetation on these topographic maps. If stereo-observation is used, this requirement may be relaxed to 5 m pixels.

For non-European mapping standards, where individual buildings and drainage need not be shown, it will be sufficient to use 5 m pixels with monoscopic observation, or 20m pixels with stereoscopic observation. It is for this reason that we, in 1977, suggested to the Government of the Federal Republic of Germany the inclusion in the first mission of Spacelab of a photogrammetric camera experiment aiming at a photographic resolution of about 20m, which is the equivalent of about an 8 m pixel size.

The first Spacelab mission was prepared jointly by ESA (European Space Agency) and NASA of the United States of America. The Federal Republic of Germany funded the camera experiment. Consequently, project management was in the hands of the DFVLR, cartographic input was entrusted to the University of Hannover and the integration of the experiment into Spacelab, as well as its representation to NASA was handled by the European Space Agency.

The first Spacelab mission was launched on NASA's Space Shuttle flight number 9 from Cape Canaveral on November 28, 1983. The shuttle landed 9 days later, on December 7, in Dryden, California. The shuttle carried the European built Spacelab, which was suitable for work for five astronauts.

The pressurized laboratory and its open cargo bay carried 37 major experiments, the photogrammetric camera experiment being number 33. The hardware consisted of an aerial mapping camera (RMK 30/23 from Zeiss Oberkochen) mounted above the Spacelab window. The camera operated from the 250 km orbit used two film cassettes, the first one containing Kodak 2443 false colour infrared film, and the second Kodak Double-X black and white film. The exposure of the film was controlled by computer.

Five hundred and fifty colour infrared and 470 black and white images were exposed, at the image scale of 1:820 000. An area of about 13 million square kilometres was covered in single strips of 190 km width, taken mostly at 60% overlap, but in mountainous areas an 80% overlap was employed for increased height precision.

About 10 million square kilometres of the imagery is cloud free, covering parts of China, the Middle East, Africa and Europe. Parts of North, Central and South American were also covered. While the Government of the Federal Republic of Germany provided the funding for the experiment, the European Space Agency determined the areas to be photographed by the way of an international call for proposals. A total of 108 requests were obtained from mapping and resource agencies all over the world. Unfortunately, due to time-line restrictions imposed by other 36 Spacelab experiments, not all requests could be satisfied.

A few examples will demonstrate the type of imagery obtained from the experiment: one of the first was a strip ranging from the Ganges Valley over Mount Everest into China. The highlands of Tibet with their frozen lakes look particularly interesting. The images from China include coverage of the Gobi desert. Unfortunately, due to the delayed launch date, images of the northern hemisphere had to be taken at a very low sun angle, since it was by then winter. However, for showing relief in Somalia, this timing was an advantage. Vegetation, appearing red on the infrared film, is well pronounced in the irrigated areas of the Al Gezirah in the Sudan, where the Blue Nile irrigates the land close to the White Nile on the left. Another image shows the field patterns of Holland and northern Germany. Even the area of Morelia in Mexico was photographed.

The more sensitive black and white film was exposed last, when solar illumination had dropped from 24° solar altitude to less than 15°. The images had to be partially underexposed and overdeveloped, which affected the expected higher black and white resolution somewhat.

After a film transport problem, which was dramatically repaired, images were taken over Western Europe where snow had fallen, making for good contrast in the images from the Alps of Austria and Germany.

Magnification of this imagery to a scale of 1:100 000 shows the street patterns of Munich, and individual buildings can be recognized, such as the German Museum and the Octoberfest area.

All 1020 images obtained have been reproduced on a microfiche catalogue. This catalogue and first generation diapositives can be ordered from the German Space Agency DFVLR in Oberpfaffenhofen at their reproduction cost. These images are now being evaluated for their mapping capacity.

First we analysed the resolution. With some images, we took edge traces of lakes with a microdensitometer. When analysing these traces, a resolution of 40 lp/mm was obtained for black and white, and for colour infrared images of estimated contrasts of 2:1. This corresponds to 20m topographic resolution or a pixel equivalent of 8 m.

As a measuring and mapping instrument an analytical plotter (Planicomp C100 from Zeiss, Oberkochen) was used. Bundle block aerial triangulation adjustments and single model evaluations with this instrument yielded point accuracies expressed in standard deviations of about 10 m in planimetry and 20 m in elevation. Investigations of the capability for line mapping are not yet completed, but experiments were undertaken for photomapping. The instrument for this purpose was the Z2 Orthoprinter by Zeiss, Oberkochen.

For differential rectification with this instrument, a digital terrain model had to be measured with the Planicomp; such a DMT is also obtainable by calculation from available digitized contours. Here the photogrammetrically measured DTM was utilized to derive contours by calculation. A position-correct orthophoto for the mountainous area of Innsbruck was derived on the basis of the DTM. Other orthophotos were derived in colour for an area near Bremen in northern Germany.

Whereas previously about six images were required to derive one orthophotomapsheet at the scale 1:1 000 000, it is now possible to derive six orthophotosheets from one image — a considerable saving of time.

A superimposition of an orthophoto on an existing Sudanese map revealed large discrepancies of locality in available African maps.

Spacelab images are also suitable for evaulation by digital image processing. For this purpose, the images need to be digitized by a drum scanner, e.g. the Optronics P1700. If this is done in three colours, the digitized image is suitable for multispectral classification, as has been done for the AL Gezirah area. Alternatively, a Spacelab black and white image may be superimposed in a colour additive manner in red with a Landsat RBV and a Landsat MSS image in blue and green. Since the images were taken at different times, colour represents change.

A comparison between an existing topographic map 1:100 000 and a Spacelab image may finally be made by a colour-space transformation of both images and their ratio into hue, intensity and saturation. This depicts huge differences of generalization in the city area for the map of 1:100 000 of Munich.

Because of bad illumination conditions during the original EOM-1 mission in July 1985, NASA has promised a reflight of the experiment. For the second mission, image motion compensation will be used, whereby the film is moved during exposure, thus compensating for the forward movement of the satellite. Less sensitive, higher resolution films will be used, which is expected to raise the resolution by a factor of 2 to 3. In this way a resolution with a pixel equivalent of 3 m may be obtained, which is likely to satisfy the requirements for 1:50 000 topographic mapping.

In the U.S.A., NASA is preparing a similar experiment with a large format camera. This camera, produced by Itek Inc. of the U.S.A. has an image format of 24 × 48 cm and a focal length of 30 cm and operates from a container mounted on an unmanned platform. It has been tested by the USGS with image motion compensation and low sensitive, high resolution films. Results indicate that this camera will be compatible in resolution with the metric camera. Hopefully, it will return from its launch this week with another 2000 high resolution photographs from space suitable for topographic mapping at medium scales.

As one Soviet Russian colleague put it to me recently: "I don't understand why such a facility is not being used every time the Space Shuttle goes up". He said: "We are doing this with good success with the MKF 6 camera, obtaining images 10m photographic of 4m pixel size resolutions any time a Sojus-Saljut mission takes place".

While other satellite systems, such as the Thematic Mapper of the anticipated SPOT, will be able to satisfy the demands for repeated image coverage and object classification better, image acquisition by electro-optical systems is now, and most likely will be in the future, limited cost-wise to a data transmission rate of 200 Mbit/s leading to a 10 m pixel size in one colour or 20 m pixel size in several colours.

Photographic images of a pixel equivalent up to 3 m have the added advantage that they may be evaluated by standard photogrammetric equipment, which is already available world-wide.

In conclusion, it is hoped that photography from space will remain as a cost-effective alternative to aerial mapping, which can be optimally used in combination with other types of satellite imagery.

8

Thematic Mapping of Natural Resources with the Modular Optoelectronic Multispectral Scanner (MOMS)

*H. Bodechtel**

ABSTRACT

This paper summarizes the main applications of remote sensing data to the development of natural resources in connection with flights of the Modular Optoelectronic Multispectral Scanner (MOMS), and discusses the technical performance of the MOMS, the evaluation of its data and its future use.

1. INTRODUCTION

Based on the Landsat system and simulation of future satellite systems, the MOMS experiment demonstrated the wide range of application for satellite remote sensing and also provided a test case of its actual performance, in some areas of application.

At present, satellite data are being used in the following geoscientific disciplines:

- geology, soils, mineral prospecting;
- forestry, agriculture, land use;
- urban and regional planning.

* Address of the author: Professor Dr H. Bodechtel, Ludwig-Maximillians-Universität, Institut für Allgemeine and Angewandte Geologie, D-8000 München 2, Federal Republic of Germany.

Past successes in the field of remote sensing have led to the definition of a new generation of earth observation satellites. The realization of the Thematic Mapper on Landsat 4, enabled the U.S.A to offer multispectral satellite images with a ground pixel resolution of 30 m and 7 spectral bands (0.4 – 12 μm), and optimized possibilities for the use of remote sensing in the fields of exploration for economic mineral resources, agricultural land use and hydrology.

In 1985, France is planning to launch a satellite system with SPOT (Système Probatoire de l'Observation de la Terre), designed as an operational monitoring system based on the new pushbroom technology. Particularly noteworthy is its high spatial resolution featuring 20 m multispectral and 10 m panchromatic. Also, SPOT is the first scanning system with a track-by-track stereo imaging mode.

New applications for a land observing system were demonstrated by the Shuttle Imaging Radar (SIR-A), to be carried on by SIR-B in 1985.

The first European satellite system to be designed for oceanographic applications was the ERS-1 of ESA. Land observation missions are being planned for 1989/90. Until now, oceanographic remote sensing data have been obtained mainly from weather satellites, e.g. the geosynchronous GEOS, Meteosat and GMS or the polar orbiting NIMBUS and SEASAT systems.

The Japanese LOS programme constitutes yet another land observation system, utilizing the new CC-technology, as do MOMS and SPOT.

A contribution from the Federal Republic of Germany is the Metric Camera onboard the Spacelab/STS-9, with cartography as its main field of application.

The scenario of optical remote sensors is being supplemented by flexible modular systems like the MOMS, geared towards specific tasks. Thanks to its modular structure, MOMS can accommodate a wide variety of user requirements.

The main features to be developed in a spaceborne earth observation system are:

- smaller ground pixel size, i.e. 20 – 30 m;
- higher spectral sensitivity;
- application-optimized band centre frequencies and band widths;

- extension of frequencies registered towards the shortwave reflected infrared range between 1.6 and 2.2 μm and emitted infrared spectrum between 8–12 μm;
- stereoscopic data acquisition;
- active SAR remote sensing.

2. DIGITAL IMAGE PROCESSING AND INTERPRETATION TECHNIQUES IN REMOTE SENSING FOR EARTH RESOURCES

Processing techniques, image enhancement and digital classification are the tools for interpreting remote sensing data.

As a first step, the information which is initially stored as row data on CCT (Computer Compatible Tape) has to be preprocessed. Geometric and radiometric corrections are necessary, as well as special corrections to compensate for haze, data transfer and variations in illumination, caused by atmospheric scattering and seasonal conditions related to the date of launch.

As a second step, single band processing is applied to the data. Linear and non-linear contrast and edge enhancement techniques (filtering) are applied to both structures and signatures in order to achieve an optimum starting point for the interpreter.

Multiband processing, including rationing and principal component analysis, is a well-known technique and commonly used to enhance spectral signatures by simultaneous elimination of relief effects.

However, in order to be effective, image data must be displayed so as to provide information in an extractable form and be optimized towards the particular need of the user. It is now possible to merge single or multiband processed multispectral-, temporal-multisensor derived data in such a way that skilled photointerpreters will be able to utilize image attributes, such as signature, texture pattern, size, shape, temperature, roughness and other physical parameters.

A common additive false colour composite, RGB (red, green, blue) shows various information given by texture, due to the ground resolution, and presents spectral differences in the processed data-bands expressed by distinct colours.

Thanks to a method using colour, defined in terms of

intensity, hue and saturation (IHS), it is possible to combine in a meaningful way data from different sensors and relating to distinct physical content.

Interesting results can be also obtained by merging different sets of data via a synthetic stereo effect.

3. MINERAL RESOURCES AND GEOLOGY

Remote sensing by satellite and digital processing of data have become indispensable tools in applied geology. A broad range of applications are possible, e.g. geological mapping, techtonical mapping, hydrocarbon exploration, exploration for geothermal resources and mineral deposits, geomorphological mapping and vegetation mapping.

Geological phenomena need to be investigated in an integrated survey in which remote sensing plays an important and basic role. Table 8.1 summarizes the main requirements of an imaging satellite system for the investigation of mineral resources, while Table 8.2 gives the identification of relevant phenomena.

Table 8.1 Mineral resources requirements.

No Monitoring Required
Large areal coverage indispensable
Multisensor approach indispensable
Data acquisition under certain and differing seasonal, meteorogical and climatological conditions
Multitemporal and repeating coverage required *e.g.* to trace subsurface geology dependent indicators, example: vegetation stress

Table 8.2 Identification of relevant phenomena.

Direct identification or classification of lithological, petrographic or mineralization units only under special conditions possible
Direct identification of interpretation of structural phenomena optimal Identification and interpretation of geological units due to *tracers* or *indicators*
e.g. vegetation moisture (soil)

4. SATELLITE REMOTE SENSING APPLIED TO LAND USE (AGRICULTURE AND FORESTRY)

Management of the environment is one of the most urgent challenges of our time. With rising consumption of renewable resources, such as cereals and wood, it becomes increasingly necessary to survey existing cultivated areas as well as to identify new regions that might be suitable for cultivation. Both tasks can be furthered by remote sensing techniques, with the following specific objectives:

- improve the scientific understanding of the environment (both biological and man-made processes);
- develop and promote practices, economic and commercial, to better handle available resources; and
- promote and develop applications of remote sensing techniques (including practical experiments).

Expected output from advanced satellite systems applied to land use (including agriculture and forestry) are:

Agriculture (see also Table 8.3)
- early determination of average for the main crops;
- automatic inventories and their regular updating;
- standardization of statistics internationally;
- yield forecast for cereals;

Table 8.3 Requirements for the application of satellite remote sensing in agriculture and land use.

Ground resolution	10 – 30 m
Repetition rate	minimum 4 images according to vegetation periods
Spectral bands	5 Bands between 0.4 and 1.1 μm 1.55 – 1.75 μm 2 bands between 8 and 12 μm synthetic aperture radar
Geometric accuracy	1 pixel
Desired scales of thematic products	1 : 50 000 – to 1 : 100 000
Availablity of data	realtime

- improved utilization of soils and selection of most suitable crops;
- quantification of disasters (natural and resulting from human error);
- evaluation of effects of crop stresses.

Forestry

- forestry mapping;
- monitoring of forest health;
- estimation of timber volume;
- harvesting of timber; and
- monitoring clear cuts.

Land use

- land use mapping at medium scales;
- land use classification;
- monitoring of dynamic land use parameters;
- monitoring of ecological and environmental parameters;
- regional and urban planning;
- real time land use data banks.

Remote sensing can be extremely useful in solving basic problems, especially in developing countries. Following are some typical occurrences that may threaten the environment and some measures to rehabilitate it:

- decrease of forests;
- decrease in farm land;
- over-grazing;
- desertification;
- salinization;
- erosion;
- vegetation damages;
- search for new ground water reservoirs;
- determination of areas of potential flooding;
- recultivation of forests;
- planning and extension of infrastructure.

Arid zones constitute about 26% of the world's land surface. In a time of overpopulation — especially in developing countries, many of which lie in arid or semi-arid zones — these dry regions cannot be excluded from environmental planning.

Because of their sparse vegetation, arid zones are ideally suited for the application of remote sensing techniques (see Table 8.4).

Table 8.4 Remote sensing technique and their application in arid zones.

	Remote sensing detection
Desertification	
Natural factors	
medium-interval precipitation decrease	multitemp. (yearly) near i.r.-data digital processing (growth rate)
vegetation changes	multitemp. (yearly), i.r. and vis. satellite data, i.r.-aerial photos
soil degradation	multitemp. (seasonal), multispectr. high-resolution data (TM, MOMS, SPOT, aerial photos)
land erosion	multitemp. MSS-data and aerial photos
anthropogenic factors	
excessive cultivation	i.r., high-resolution: TM, MOMS, aerial photos
overgrazing	
fire clearing/wood cutting	*vis*, MSS and aerial photos
Salinization	
base rock	MSS, digital processing surface roughness by microwave measurement
soil cover	MSS (*vis* to microwave)
subsurface water resources	soil moisture by L-band radar thermal i.r. lineament analyses on MSS data and aerial photos
drainage system	*vis* aerial photos high-resolution satellite data
water chemistry	near i.r., therm. i.r.
Water resources developing and irrigation	
technical planning	high resolution, *vis*, satellite data
borehole siting	aerial photos (lineaments, aquifer mapping) near i.r. (vegetation distribution) Landsat MSS (rock type discrimination, aquifer mapping, small-scaled lineaments)
irrigation supervision	aerial i.r. photos microwave (soil moisture) *vis* and near i.r.

The total area classified as arid is increasing worldwide. Each year thousands of km^2 of previously semi-arid land becomes arid. This is partially a result of misuse, such as overgrazing, vanishing forests by firelacing and wood cutting, and erosion-causing methods of cultivation.

With satellite imagery, areas affected by desertification may be mapped, the data collected elaborated in combination with other remote sensing information, which may lead eventually to the identification and protection of other areas similarly threatened. Moreover, salinization of cultivated plains can be clearly measured in terms of grade and extension by measuring their relatively high albedo. Not all arid zones are useless for cultivation. With satellite imagery, agroecological and environmental planning may fruitfully be applied to apparently arid regions.

Satellite data evaluation may also serve as a basis for irrigation projects and surveys. For example, lineament density analyses of groundwater development in old crystalline rocks provides first-hand information for the determination of future well-drilling points.

Groundwater storage capacity may be estimated either from secondary indicators, such as variations in vegetation, differences in spectral signatures, airborne geoelectrics measurements, or directly from spaceborne or airborne microwave soil moisture and surface roughness measurements.

Investigations of alluvial grid patterns revealing a high vegetation growth rate may suggest new areas suitable for cultivation.

5. *THE MISSION OF THE MODULAR OPTOELECTRONIC MULTISPECTRAL SCANNER (MOMS)*

The Modular Optoelectronic Multispectral Scanner (MOMS) was sponsored by the Federal Ministry of Research and Technology (BMFT), with Messerschmitt-Bölkow-Blohm (MBB) as the main contractor responsible for its development, under contract from the German Aerospace Research Establishment (DFVLR) and under the scientific supervision of the University of Munich.

This new instrument, designed for regional and global

optical remote sensing, may be flown either on aircraft or space platforms. In two space missions, MOMS was mounted on the Shuttle Pallet Satellite (SPAS): aboard shuttle flights STS-7 in June 1983 and STS-11 in February 1984. The missions served the dual purpose of verifying the technical operation of the sensor in space, and conducting geoscientific and application oriented experiments around the globe.

5.1 MOMS Technical Performance

MOMS-01 for the first time makes use of CCD technology by means of the push-broom principle (Fig. 8.1). Scan line extension beyond 1 CCD array (up to 6 array per focal plane feasible) is done by the dual optics principle.

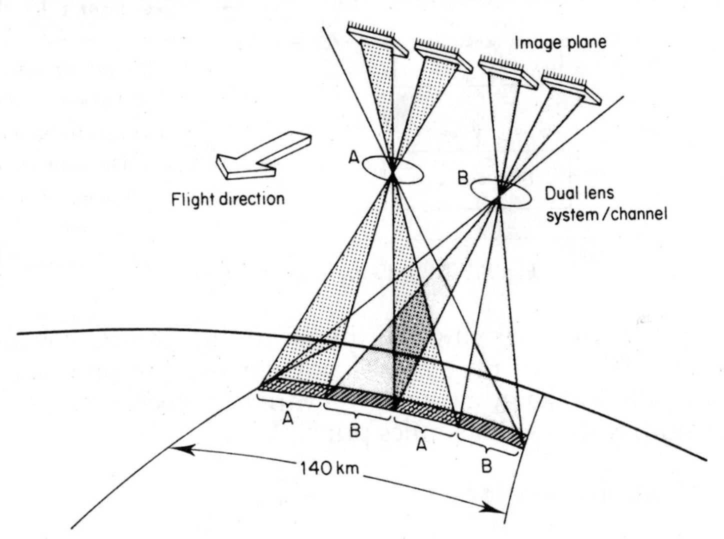

Fig. 8.1. MOMS-01 CCD-Scanning principle, scan lines are extended with the dual lens system.

The most useful characteristic of the MOMS is its modular structure, the individual modules being: filters, optics, sensors and preamplifier electronics allocated to specific spectral bands (see Fig. 8.2).

This modularity allows the MOMS to be adapted to completely different types of mission, e.g. land surface thematic mapping, sea or vegetation monitoring, coastal zone evaluation and, in a stereo mode, conventional photo interpretation

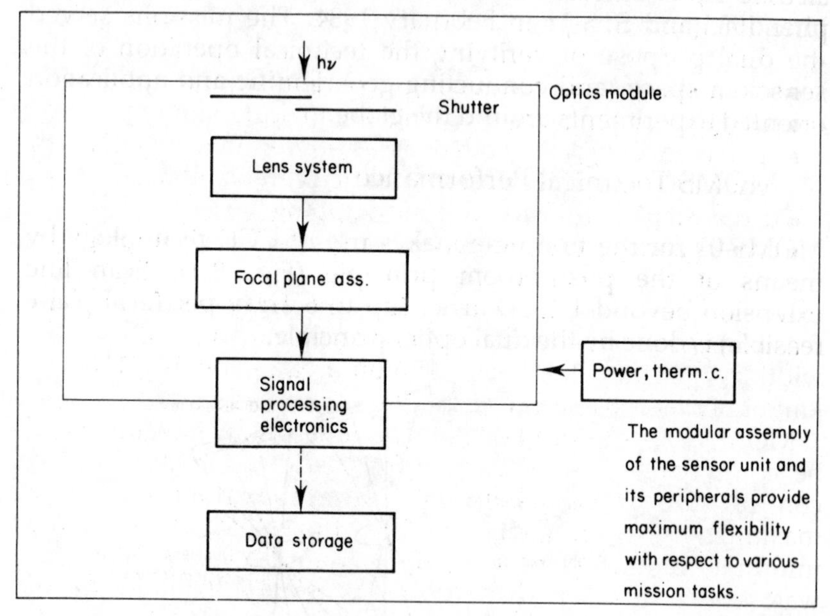

Fig. 8.2. MOMS modular concept.

and topographic mapping. Furthermore, the ground coverage may be adapted to user needs by changing the photometric layout and focal length, combining various detector arrays (line extension) with the dual optics principle.

5.2 MOMS-01 Missions

The MOMS-01 missions added a number of remarkable events to earth observation remote sensing:

- first modular system successfully tested in space;
- first high resolution (20 m) earth observation sensor flown in space using CCD (Charge Coupled Devices) technology;
- first remote sensing system to have been refurbished and reflown in space within a short period of time.

In order to demonstrate the capability of the MOMS to handle two-channel high-resolution CCD data for thematic mapping up to scale 1 : 50 000 applicable to a broad spectrum of geoscientific disciplines, it was desirable to select sites from all over the world, thus obtaining a large variety of geoscientific phenomena for imaging:

- arid and semi-arid areas — for geological interpretation;
- mountainous regions — for morphological mapping;
- sparse to dense vegetated areas — for investigations of renewable resources (forestry, agriculture, land use);
- urban areas — for cartographic mapping;
- coastal zones and open ocean islands.

Due to the experimental character of the missions, the recording time was limited to about 30 min. Given a swath width of 140 km, this is equal to an area of roughly 1800000 km^2 or nearly 150 image frames.

MOMS status information and commands for imaging could be controlled during the missions, either from the mission control center at JSC in Houston or from aboard the shuttle. To maximize the scientific benefit from the MOMS missions, an information network for a real-time controlled data acquisition was set up. In order to identify areas most suitable for image sequences, geoscientific information and weather data from geostationary weather satellites were collected one and a half hours before the actual data take. This allowed the MOMS-01 to record a total number of 26 image frames under mostly optimum weather conditions.

The areas imaged and recorded lie between latitudes 28 north and 28 south consequent to the orbit inclination of the STS-7 and STS-11 flights, as shown in Tables 8.5 and 8.6.

5.3 MOMS Data Evaluation

Processing and archiving of MOMS data was conducted at the German Remote Sensing Data Center at Oberpfaffenhofen. To generate user products, the computation and photolab facilities will use different processing steps, such as:

- quick look from HDT (high density tape);
- conversion from HDT to raw data CCT (Computer-Compatible Tape);

Table 8.5 STS-7 data takes.

Serial no.	Orbit	Area	Longitude (deg) from	Longitude (deg) to	Length (S)	No of Scenes	Sun elev (deg) from	Sun elev (deg) to	Comment
1	37/1	Chile Bolivia	-70.5	-63.0	120	10	40.9	34.3	Andes profile (Arica)
2	41	Hong Kong	110	115.8	90	7	26.8	31.9	Canton Hong Kong
3	42/1	India	86.5	92.3	90	7	26.0	31.0	Ganges delta
4	60	Egypt Saudi Arabia Maldives	23.0	96.6	1290	108	10.9	57.1	Lake Nasser with Aswan Dam, Jeddah Mecca, Taif

Table 8.6 STS-11 (41-B) data takes.

Serial no.	Orbit	Area	Longitude (deg) from	Longitude (deg) to	Length (S)	No. of Scenes	Sun elev (deg) from	Sun elev (deg) to	Comment
1	20	Brazil	-46.0	-43.2	48	4	46.9	44.2	Minas Gerais
2	21	Brazil	-58.0	-55.8	36	3	36.6	34.4	Rio Paraguay Brazil
3	22	Chile Argentina	-70.7	-63.5	108	9	27.1	20.5	Anto-Fagasta
4	26	Burma China, Laos	98.0	102.6	72	6	34.3	38.7	Mekong
5	27	India	83.0	84.4	24	2	41.8	43.2	Orissa
6	28	Saudi Arabia	44.0	54.9	168	14	26.4	36.7	Riyadh
7	29	Egypt Sudan	30.0	32.3	36	3	34.7	36.9	Lake Nasser
8	29	Australia	119.5	121.8	36	3	37.6	35.5	Mining district (Stereo)
9	30	Algeria	5.0	7.3	36	3	32.6	34.8	Hoggar Mountains
10	30	Australia	115.0	116.6	24	2	20.5	19.0	Western Territory
11	47	Mali Upper Volta	-6.0	-1.2	84	7	47.2	52.2	Niger River

Serial no.	Orbit	Area	Longitude (deg) from	to	Length (S)	No. of Scenes	Sun elev (deg) from	to	Comment
12	52	Brazil	-49.0	-45.1	60	5	47.5	43.7	São Paulo
13	53	USA	156.2	-154.8	24	2	32.2	33.8	Hawaii
14	58	India	83.0	84.5	24	2	24.5	26.1	Orissa
15	62	Uganda Kenya	30.5	40.3	180	15	66.8	76.9	East Africa Rift Valley
16	63	Senegal Guinea Ivory Coast	-16.9	2.1	336	28	38.6	60.4	White Nile
17	75	Australia	136.0	138.3	36	3	63.9	61.6	Northern Territory
18	76	Sudan Yemen	34.6	47.5	216	18	23.4	37.7	Sana
19	76	Australia	115.0	121.9	108	9	62.3	55.2	Western Territory (Stereo)
20	77	Ethiopia	37.5	40.8	60	5	52.5	56.5	Rift Valley
21	78	Niger	10.0	44.0	60	5	36.4	40.3	Niamey
22	84	Brazil	-47.9	-47.4	7	½	44.8	44.3	Rio Parana

- radiometric and geometric correction from raw CCT to corrected CCT;
- generation of high quality images on film and photolab processing.

Results are the following user products:

- quick look images;
- CCTs on standard CCT format 1600 Bpi (raw data, corrected data);
- black and white images as transparencies and paper prints.

Aside from purely geoscientific thematic interpretation of the MOMS data, another aim will be the comparative and complementary analysis of MOMS-01 and Thematic Mapper data on the basis of a BMFT-NASA mutual data exchange agreement.

A major effect will be put into the following topics:

- comparative investigations of MOMS and Thematic Mapper data concerning the differentiation and identification of geoscientific phenomena;
- merging of MOMS data with multisensor data to examine their applicability to certain geoscientific problems;
- discussion of the advantages of high spectral and spatial resolution for various application areas to optimize these parameters for future MOMS instrument developments.

5.4 Future MOMS Uses

MOMS-01 is the first space-qualified representative of a family of instruments, developed in an effort to create a complex remote sensing device adaptable to a variety of configurations.

Further developments, already carried up to design status, include:

- refurbishment of the existing MOMS-01 instrument with one or two additional spectral bands in the visible and near infrared range to obtain more complex spectral information for geoscientific thematic mapping;
- a Stereo-MOMS instrument providing *in situ* panchromatic stereoscopic coverage with 10 m base and less than 15 m height resolution.

Main applications will include:

- topographic mapping 1:25 000 — 1:50 000;
- evaluation of digital terrain model;
- orthophoto maps;
- optimized geoscientific thematic mapping 1:250 000 — 1: 50 000 on the basis of combined spectral and topographic data;
- quantification of multispectral data under consideration of relief dependent parameters;
- a SWIR-MOMS module with one or two spectral bands in the short wave infrared range at 1.6 and 2.2 μm.

Main applications are:

- improved lithological mapping or exploration of mineral/ fossil resources;
- observations/measurements of vegetation;
- general thematic interpretation;
- measurements of soil and vegetation moisture.

From the point of view of application, the MOMS-programme — but not necessarily the MOMS technology — is focussed on Shuttle related activities. A multispectral MOMS instrument flown in connection with optical (metric camera, large format camera) or active microwave sensors (SIR, Radarsat) could provide complementary spectral information.

Unlike Landsat and SPOT, the configuration of the MOMS can be changed both during and between Shuttle missions. To achieve the latter, a retrievable platform such as SPAS or EURECA (developed by ESA) will have to be planned for. These facts should be borne in mind when designing any MOMS mission.

A particular mission may serve either a purely scientific objective, or an application oriented one, or both.

The combination of the multiple alternatives offered by the great number of Shuttle orbits and the modular structure of the MOMS, giving it a high level of flexibility and adaptability, is considered to provide optimum opportunities for data acquisition in connection with existing and planned remote sensing satellites.

9

Availability of Remotely Sensed Data and Information from the U.S. National Oceanic and Atmospheric Administration's Satellite Data Services Division

*Bruce H. Needham**

ABSTRACT

The chapter describes the main satellites which provide data bases for meteorological and oceanographic monitoring and resources development. Background or geostationary and polar orbiting satellites as well as the sensors from aboard those satellites is documented in order to guide in the use of products and services provided by NOAA.

1. INTRODUCTION

The Satellite Data Services Division (SDSD), part of the U.S. National Oceanic and Atmospheric Administration's (NOAA) National Environmental Satellite, Data, and Information Service (NEDIS), acts as the official U.S. archive of all data and products from NOAA's operational geostationary and polar orbiting satellites, and several of the National Aeronautics and Space Administration's (NASA) experimental satellites.

The SDSD was formed in 1974 when NOAA realized that the need existed to maintain a formal, organized archive for all the data and products from its operational environmental satel-

* Address of the author: Bruce H. Needham, Chief of the Data Services Branch, NOAA, Satellite Data Services Division, Maryland 20233, U.S.A.

lites, and the potential use of these data and products by users throughout the world on a retrospective basis.

The SDSD archive of data and products dates back to April 1960 and contains data from then to the present time. It is a unique source of data and information relevant to many scientific diciplines. While primarily intended for meteorological purposes, many of the sensors carried on the more recent spacecraft also provide data of great value to oceanographers, hydrologists, geologists, agronomists and others.

The SDSD is collocated with the operations center of NESDIS which manages the National Operational Environmental Satellite Program. This collocation expedites SDSD acquisition of the environmental data. It also allows SDSD personnel to monitor the latest satellite data applications, to note outstanding environmental events which may be of interest to subsequent users, and to ensure that the original imagery negatives and magnetic data tapes reach the Division undamaged and as quickly and economically as possible.

The SDSD archive currently has data and information from nearly 40 satellites which has carried over 30 different instruments from 1960 up to the present. Data holdings include over 250 000 digital tapes, over 10 million photographic images in various formats and sizes, and over 5000 analysis charts.

2. SATELLITE SYSTEMS AND SENSORS

Figure 9.1 illustrates the U.S. satellites with data on archive at SDSD. These satellites, dating back to TIROS-1 (Television Infrared Operational Satellite) in 1960 up to the present NOAA-8, and GOES-6 satellites, are basically comprised of two types of spacecraft: polar orbiting and geostationary.

2.1 Polar Orbiting Satellites

Polar orbiting satellites are in a relatively low orbit around the Earth (approximately 500-900 miles, or 800-1500 km). This allows them to circle the globe from 12 to 14 times each day and obtain imagery and quantitative digital data along a path on the Earth's surface up to 1550 miles (2500 km) in width during both daytime and night-time. Table 9.1 lists the pertinent orbital characteristics and other information on these satellites.

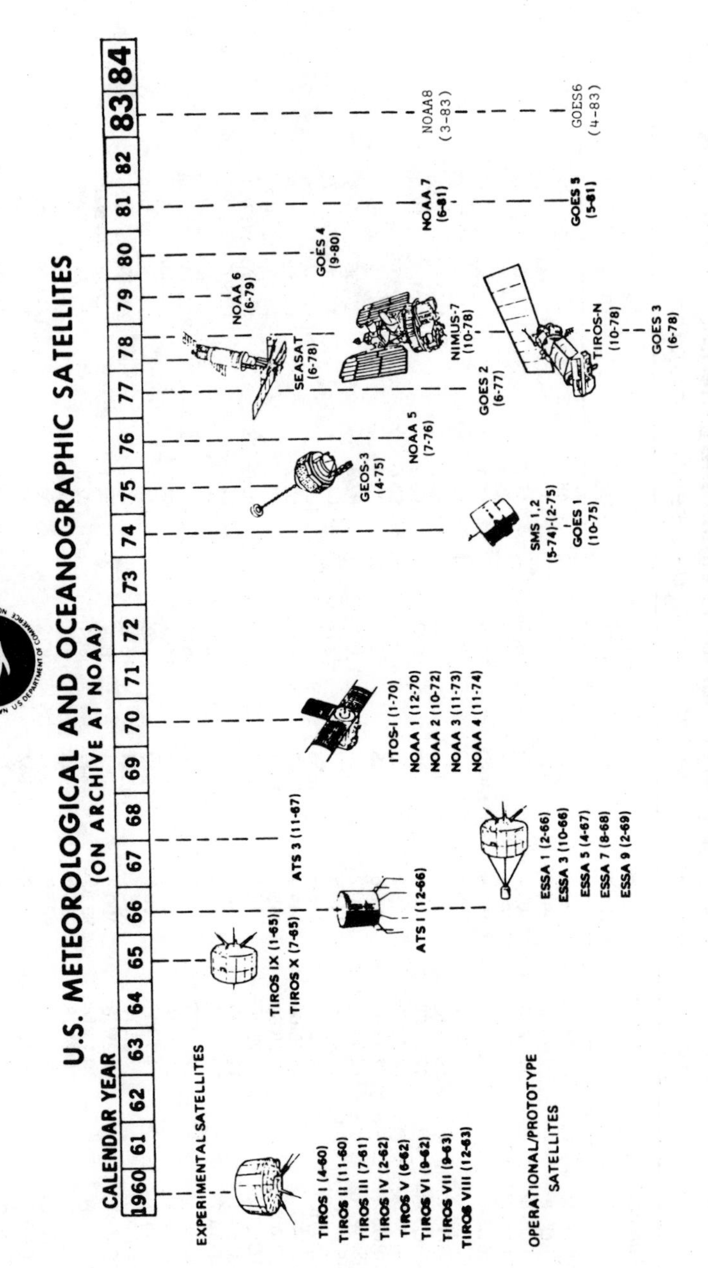

Fig. 9.1. U.S. Meteorological and oceanographic satellites.

Table 9.1. U.S. Polar-orbiting environmental satellites and experimental satellites whose data are archived at satellite data services division.

Satellite	Launched	Period (min)	Perigree (km)	Apogee (km)	Inclination (degrees)	Dates of data on archive at SDSD
TIROS-I	4/1/60	99.2	796	867	48.3	4/1/60 – 6/14/60
TIROS-II	11/23/60	98.3	717	837	48.5	11/23/60 – 9/27/61
TIROS-III	7/12/61	100.4	854	937	47.8	7/12/61 – 1/23/62
TIROS-IV	2/8/62	100.4	817	972	48.3	2/8/62 – 6/18/62
TIROS-V	6/19/62	100.5	680	1119	58.1	6/19/62 – 5/14/63
TIROS-VI	9/18/62	98.7	783	822	58.2	9/18/62 – 10/21/63
TIROS-VII	6/19/63	97.4	713	743	58.2	6/19/63 – 2/26/66
TIROS-VIII	12/21/63	99.3	796	878	58.5	12/21/63 – 2/12/66
TIROS-IX	1/22/65	119.2	806	967	96.4	1/23/65 – 9/9/66
TIROS-X	7/2/65	100.6	848	957	98.6	7/2/65 – 4/2/66
ESSA-1	2/3/66	100.2	800	965	97.9	2/4/66 – 10/6/66
ESSA-3	10/2/66	114.5	1593	1709	101.0	10/4/66 – 6/1/67
ESSA-5	4/20/67	113.5	1556	1635	101.9	6/1/67 – 12/3/68
ESSA-7	8/16/68	114.9	1646	1691	101.7	4/16/68 – 3/31/69
ESSA-9	2/26/69	115.3	1637	1730	101.9	4/1/69 – 11/15/72
ITOS-1	1/23/70	115.1	1648	1700	102.0	4/28/70 – 6/17/71
NOAA-1	12/11/70	114.8	1422	1472	102.0	4/26/71 – 6/20/71
NOAA-2	10/15/72	114.9	1451	1458	98.6	11/16/72 – 3/19/74
NOAA-3	11/6/73	116.1	1502	1512	101.9	3/26/74 – 12/17/74
NOAA-4	11/15/74	101.6	1447	1461	114.7	12/17/74 – 9/15/76
GEOS-3	4/9/75	100.6	840	848	115.7	4/14/75 – 12/1/78
NOAA-5	7/29/76	116.2	1504	1518	102.1	9/15/76 – 3/16/78
SEASAT	6/27/78	100.8	794	808	108.1	7/7/78 – 10/9/78
TIROS-N	10/13/78	98.92	849	864	102.3	2/5/79 – 11/1/80
NIMBUS-7	10/24/78	99.28	943	955	104.9	10/24/78 – present
NOAA-6	6/27/79	98.68	805	822	98.7	6/27/79 – 6/20/83
NOAA-7	6/23/81	102.04	852	870	98.9	6/23/81 – present
NOAA-8	3/23/81	101.29	805	822	98.7	6/20/83 – 6/12/84

ESSA-2, -4, -6, and -8 were automatic picture transmission (APT) satellites; hence no negatives are archived.
NIMBUS-7 — Only data from the coastal zone color scanner (CZCS) are archived at SDSD.

Table 9.2. Polar orbiter NOAA satellite sensor characteristics.

Satellite (series)	Sensor name	Band	Spectral range	Ground resolution	Swath width
TIROS-I through X	Vidicon	VIS	0.45 – 0.65 μm	3.8	800 (km)
ESSA 1, 3, 5, 7, 9	Advanced vidicon camera systems (AVCS)	VIS	0.45 – 0.65 μm	2.2	2300
	low resolution infrared radiometer (LRIR)	i.r.	———	———	variable
ITOS-1/NOAA-1	AVCS	VIS	0.45 – 0.65 μm	2.2	2300
	scanning radiometer (SR)	VIS	0.50 – 1.00 μm	4	4000
		i.r.	10.50 – 0.65 μm	8	4000
NOAA-2 through 5	scanning radiometer (SR)	VIS	0.50 – 12.50 μm	4	4000
		i.r.	10.50 – 12.50 μm	8	4000
	very high resolution radiometer (VHRR)	VIS	0.60 – 0.70 μm	0.8	2580
		i.r.	10.50 – 12.50 μm	0.8	2580
	vertical temperature profile radiometer (VTPR)	1	14.92 – 14.99 μm	68	1876
		2	14.65 – 14.87 μm	68	1876
		3	14.29 – 14.49 μm	68	1876
		4	14.03 – 14.22 μm	68	1876
		5	13.70 – 13.88 μm	68	1876
		6	13.30 – 13.48 μm	68	1876
		7	18.38 – 19.01 μm	68	1876
		8	11.93 – 12.08 μm	68	1876
TIROS-N/NOAA-6	advanced VHRR (AVHRR)	1	0.550 – 0.90 μm	1.1-4	2580/4000[a]
		2	0.725 – 1.10 μm	1.1-4	2580/4000[a]
		3	3.550 – 3.93 μm	1.1-4	2580/4000[a]
		4	10.050 – 11.05 μm	1.1-4	2580/4000[a]
	TIROS-N operational vertical sounder (TOVS)				
	high res i.r. spectrometer (HIRS)	1-20	0.660 – 14.98 μm	17.4	1120
	stratospheric sounding unit (SSU)	1-3	14.970 μm	147.3	1473
	microwave sounding unit (MSU)	1-4	50.300 – 57.95 GHz	109	2320
NOAA-7 & 8	(same as above with addition of extra channel in AVHRR)	4	10.300 – 11.30 μm	1.1-4	2580/4000[a]
		5	11.500 – 12.50 μm	1.1-4	2580/4000[a]

[a]AVHRR resolution is 1.1 km in high resolution mode; 4km in low resolution mode. Swath width is 2580 km high resolution mode; 4000 km in low resolution.

Table 9.3 NASA Experimental satellite sensor characteristics.

Satellite (series)	Sensor name	Band	Spectral range	Ground resolution	Swath width (km)
GEOS-3	altimeter (ALT)	1	13.900 GHz (2.16 cm)	3.6 km (HR) 14.3 km (LR)	3 (HR) 9 (LR)
SEASAT	altimeter (ALT)	1	13.900 GHz (2.16 cm)	2.4 – 12 km	2.4 — 12
	scatterometer (SASS)	1	14.599 GHz (2.05 cm)	50 km	1000
	scanning microwave radiometer (SMMR)	1	6.600 GHz (4.50 cm)	136 × 89 km	600
		2	10.700 GHz (2.80 cm)	87 × 58 km	600
		3	18.000 GHz (1.70 cm)	57 × 35 km	600
		4	21.000 GHz (1.40 cm)	44 × 29 km	600
		5	37.000 GHz (0.80 cm)	28 × 18 km	600
	visible infrared radiometer (VIR)	1	0.470 – 0.94 μm	2 km	2280
		2	10.500 – 12.50 μm	4 km	2280
	synthetic aperture radar (SAR)	1	1.350 GHz (22 cm)	25 km	100
NIMBUS-7	coastal zone color scanner (CZCS)	1	0.433 – 0.453 μm	825 m	1600
		2	0.510 – 0.530 μm	825 m	1600
		3	0.540 – 0.560 μm	825 m	1600
		4	0.660 – 0.680 μm	825 m	1600
		5	0.700 – 0.800 μm	825 m	1600
		6	10.500 – 12.500 μm	825 m	1600

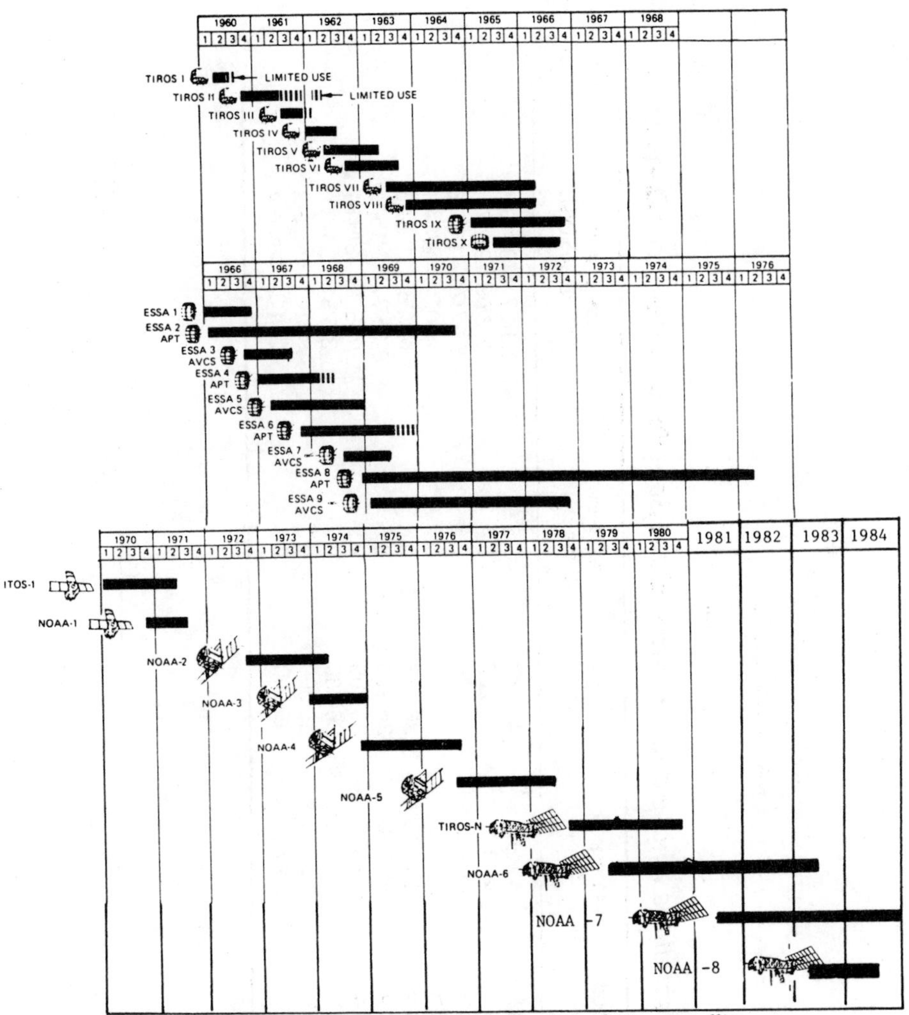

Fig. 9.2. Period of record — NOAA polar orbiting satellites.

Table 9.2 lists the sensors that each of these NOAA satellites carried, along with their spectral characteristics. Table 9.3 lists similar information from NASA experimental satellites whose data are archived at the SDSD (GEOS-3, SEASAT and NIMBUS-7). Figure 9.2 illustrates the life-span of each of these satellites in graphic form.

During normal operations, NOAA maintains two polar orbiting satellites in operation, with slightly different equatorial crossing times to permit coverage of any one spot on the Earth's surface at least twice daily for each satellite, for a total coverage of four times daily.

Table 9.4 U.S. geostationary satellites whose data are archived at Satellite Data Services Division.

Satellite	Launched	Period	Perigree (km)	Apogee (km)	Inclination (degrees)	Dates of data on archive at SDSD
ATS-I	12/6/66	24 h	41 257	42 447	0.2	1/1/67 – 10/16/72
ATS-III	11/5/67	24 h	41 166	41 222	0.4	3/2/68 – 9/2/74
SMS-1	5/17/74	24 h	35 605	35 975	0.6	6/27/74 – 1/7/76
SMS-2	2/6/75	24 h	35 482	36 103	0.4	3/10/75 – 8/4/81
GOES-1	10/6/75	24 h	35 728	36 847	0.8	1/8/76 – 3/15/80
GOES-2	6/16/77	24 h	35 600	36 200	0.5	8/15/77 – 9/15/80
GOES-3	6/15/78	24 h	35 600	36 200	0.5	7/13/78 – 3/5/81
GOES-4	9/9/80	24 h	35 782	36 200	0.2	3/5/81 – 6/1/83
GOES-5	5/15/81	24 h	35 600	35 600	0.2	7/9/81 – present
GOES-6	4/28/83	24 h	35 790	36 000	0.1	6/1/83 – present

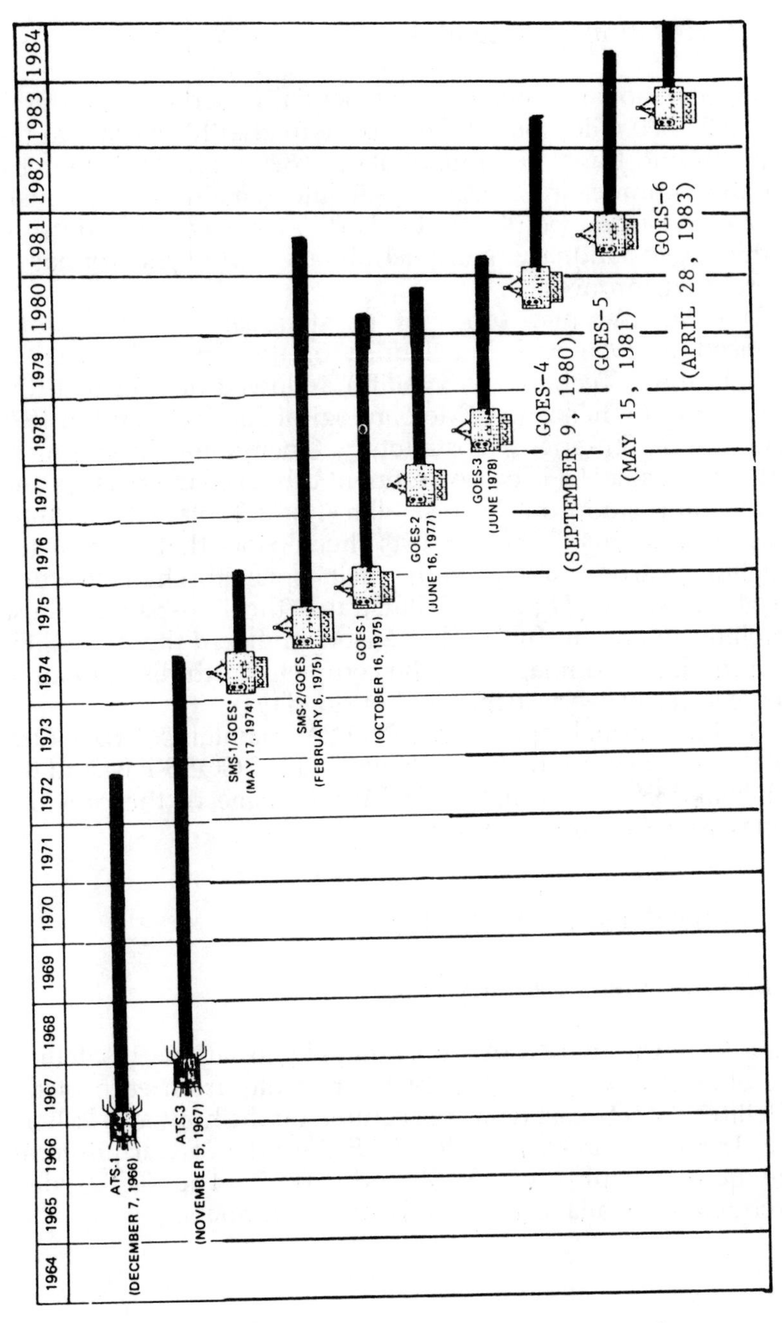

Fig. 9.3. Period of record — NOAA geostationary satellites.

2.2. Geostationary Satellites

The geostationary satellites are essentially 'parked' in an orbit about 22 000 miles (36 000 km) above the Earth's surface at the Equator and travel at a rate of about 6820 mph (11 000 km/h). At this altitude and speed, a satellite remains continuously above the same point on the Earth's surface and, thus, is termed: geostationary, geosynchronous, earth synchronous or merely synchronous.

The NOAA solar series of geostationary satellites commenced in 1966 with the launch of the first ATS satellite (Applications Technology Satellite), followed in 1974 with the SMS series (Synchronous Meteorological Satellite), and in 1975 with the GOES series (Geostationary Operational Environment Satellite). Table 9.4 lists the pertinent orbital characteristics and other information on these satellites, and Table 9.5 lists their current locations. Table 9.6 lists the sensors that each of the satellites carried along with their spectral characteristics, resolutions, etc. Figure 9.3 illustrates the life-span of these satellites in graphic form. The sensors on board these satellites acquire data and imagery of the complete Earth disk (about ¼ of the surface) every 30 min, 24 hours a day.

During normal operations, NOAA maintains two GOES satellites in operation: one at 75° west called GOES-East and the other at 135° west called GOES-West. Some of the previous series are on a stand-by mode.

3. PRODUCTS AND SERVICES

3.1 NOAA Geostationary Satellites

Data from the two NOAA operational geostationary satellites are obtained every 30 min, 24 hours a day from each of the satellites. GOES-East data are acquired on the hour and ½ hour (i.e. 1800Z and 1830Z), while GOES-West data are acquired on the quarter-hour (i.e. 1815Z and 1845Z). The GOES data products are available in several different formats.

Table 9.5. Location of geostationary satellites (of October 30, 1984).

Satellite	Operational/Standby	Location (°W)	Designation	Comments
SMS-2	standby	75		
GOES-1	standby	127		
GOES-2	standby	107		
GOES-3	standby	130		
GOES-4	standby	135		
GOES-5	non-operational	75		failed 6/16/84
GOES-6	operational	98	GOES-Mid	effective 8/4/84

GOES-5 (GOES-East) failed on 6/16/84; GOES-6 (GOES-West) was moved from its previous position at 135 West to a new position at 98 West between June 16 and August 4, 1984 to act as GOES-Mid, and will remain in this position till November 1984, at which time it will be moved to about 108 West until April or May 1985.

3.1.1. Magnetic Tape (Digital) Data

(a) *Reduced Resolution (8 km vis & i.r)*
7/26/76 to 9/5/78: nominally 5 times daily;
9/6/78 to present: nominally every 3 hours;

(b) *Full Resolution (8 km i.r., 1 km vis)*
GOES-East: 1/18/78 to present — every ½ hour;
GOES-West: 11/20/78 to present — every ½ hour;
GOES-Indian Ocean: 12/1/78 11/30/79 — every ½ hour.

(c) *Wind vector (cloud motion) fields*
GOES-East/West: 10/74 to present — compiled monthly.

3.1.2 Photographic Products

(a) *Full disk and sectors (25 × 25cm negatives) vis & i.r.*
SMS-1 (GATE): 6/27/74 — 9/21/74
GOES-East: 9/22/73 — present *
GOES-West: 3/10/75 — present *

(b) *Microfilm (full disk vis & i.r.)* days/reel
SMS-1 (GATE): 6/25/74 — 9/21/74
GOES-East: 9/1/74 — Present
GOES-West: 3/11/75 — Present

3.2. NOAA Polar Orbiting Satellites

Data from the NOAA polar orbiting satellites are also maintained as digital tapes, photographic imagery, and paper chart products.

* NOTE: GOES-East failed in the summer of 1984, and GOES-West was moved to a mid-U.S. position for an undetermined period of time.

TABLE 9.6 Geostationary NOAA satellite sensor characteristics.

Satellite (series)	Sensor name	Band	Spectral range (μm)	Ground resolution (km)	Swath width
AST-I & III	spin scan camera (SSC)	VIS	0.550 – 0.70	4	horizon to horizon
SMS-1 & 2	visible infrared spin scan radiometer (VISSR)	VIS	0.550 – 0.70	14	horizon to horizon
		i.r.	10.500 – 12.60	8	horizon to horizon
GOES 1-6	VISSR	VIS	0.550 – 0.70	14	horizon to horizon
		i.r.	10.500 – 12.60	8	horizon to horizon
	VISSR atmospheric sounder (VAS) [GOES-4 and 5 only]	V-1	0.550 – 0.70	14	horizon to horizon
		i.r.-1	14.600 – 14.81	16	horizon to horizon
		i.r.-2	14.290 – 14.62	16	horizon to horizon
		i.r.-3	14.060 – 14.39	16	horizon to horizon
		i.r.-4	13.790 – 14.18	16	horizon to horizon
		i.r.-5	13.120 – 13.48	16	horizon to horizon
		i.r.-6	4.496 – 4.537	16	horizon to horizon
		i.r.-7	12.500 – 12.82	16	horizon to horizon
		i.r.-8	10.360 – 12.12	16	horizon to horizon
		i.r.-9	7.143 – 7.353	16	horizon to horizon
		i.r.-10	6.390 – 7.067	16	horizon to horizon
		i.r.-11	4.386 – 4.484	16	horizon to horizon
		i.r.-12	3.623 – 4.310	16	horizon to horizon

3.2.1. TIROS-1 Through X Series

- 35mm microfilm and CCTs;
- 4/1/60 – 4/20/66.;

3.2.2. ESSA 1, 3, 5, 7 and 9 series

- *35mm microfilm, 25 × 25 cm negs, CCTs;*
- 2/4/66 – 11/16/71.

3.2.3. ITOS-1 Through NOAA-5, TEROS-N, and NOAA-6, 7 and 8 Series

- 4/28/70 to present;
- individual pass-by-pass (35mm, 25 × 25 cm, and CCT);
- high resolution frames (25 × 25 cm and CCT);
- polar and mercator mosaics (35mm, 25 × 25 cm and CCT);
- Sounding date (CCTs);
- SST, ice charts (charts, 25 × 25 cm, and CCTs).

Note: Interested users should contact the SDSD to ascertain the exact availability of the above. Not all data were continuously archived in all formats. Basically, only data from January 1979 to the present are available on CCT.

3.3. Non-NOAA Satellite Data

3.3.1. GEOS-3

Data from the altimeter (ALT) on GEOS-3 are archived for the period from 4/14/75 – 12/1/78 in both raw format (10 samples/s, 131 CCTs) and geophysical format (1 samples/s, 11 CCTs).

3.3.2. SEASAT

Data from the sensors on board SEASAT are archived as magnetic tapes and photographic products. Data from the ALT, SASS and SMMR are available only on CCT, global coverage, from the entire mission (7/7/78 – 10/9/78). Tables 9.7 and 9.8 summarize these products. Data from the SAR are available in 70mm photographic strips over most areas of the U.S.A. Canada, Central America, Western Europe, and adjacent waters (see Figure 9.4) and as CCTs and film products for those scenes that were digitally correlated (see Figure 9.5).

Table 9.7. SEASAT final geophysical data record content definition.

File	Record Type	Sensor			
		Alt	SASS*	SMMR	VIRR
Geophysical file	Basic geophysical record	one point time/lat/long; fully corrected h, H-1/3 and tidal height; steric anomalies; corrected SS height above ellipsoid; atm. press. effect; ionosphere correction; wet/dry tropo. orr.; surface pressure	mean time/lat/long wind stress magnitude and σ wind stress direction and σ mag/dir correlation 120 wind vector solutions with aliases; (ambiguity in direction will not be removed)	mean time/lat/long sst, ssw, rain rate; atm. liq. water/ water vapor; path length corr. 600 × 600km area; 90 seconds of data	
	Supplement geophysical record		time/lat/long; fully corrected backscatter coefficient (σ); individual SMMR channel temps (int. over area for each of 15 cells/ fan beam); 1.89 s;		

	Basic sensor record	one point; time/lat/long; instrument corr. h; fully corr. H-1/3	time/lat/long; instr. corr. back. coeff. (σ); corner lat/long; s/n ratio; σ corrections 1.89 s;	mean time/lat/long individual channel brightness temps and T_a; same area as above	start time of scan line; lat/long of start-middle-end; VIS brightness/i.r. temp. one scan line — 1.25 s.
Sensor file	Supplement sensor record	time Δ h; one minute; calibration mode	time; 7.6 min. cal. sequence noise; calibration mode	start/time/nadir lat/long cone-clock angle; indiv. chan. hot/cold cal. mean and ant. temp. footprint lat/long for 30 footprint locations one 4.096 s scan line; calibration data	start time nadir lat/long; altitude i.r./VIS hot/cold cal. calibration mode

Table 9.8. Inventory of Seasat GDR data on archive.

Sensor	Files	Dates of Data	Storage Type	# of CCTS
ALT	geophysical file	7/7-7/17, 7/24-8/28, 9/1, 9/6, 9/7, 9/10, 9/13, 9/15-10/10/78	6 days global/tape	14
ALT	sensor file	7/7-7/17, 7/24-8/28, 9/1, 9/6, 9/7, 9/10, 9/13, 9/15-10/10/78	3 days global/tape	26
SASS	geophysical and sensor file	7/7-10/10/78	6 hours global/tape	381
SASS	geophysical file	7/7-10/10/78	24 hours global/tape	96
SASS	basic geophysical record only	7/7-10/10/78	48 hours global/tape	48
SMMR	geophysical and sensor files	7/7-10/10/78	6 hours global/tape	381
SMMR	geophysical file	7/7-10/10/78	4 days global/tape	24

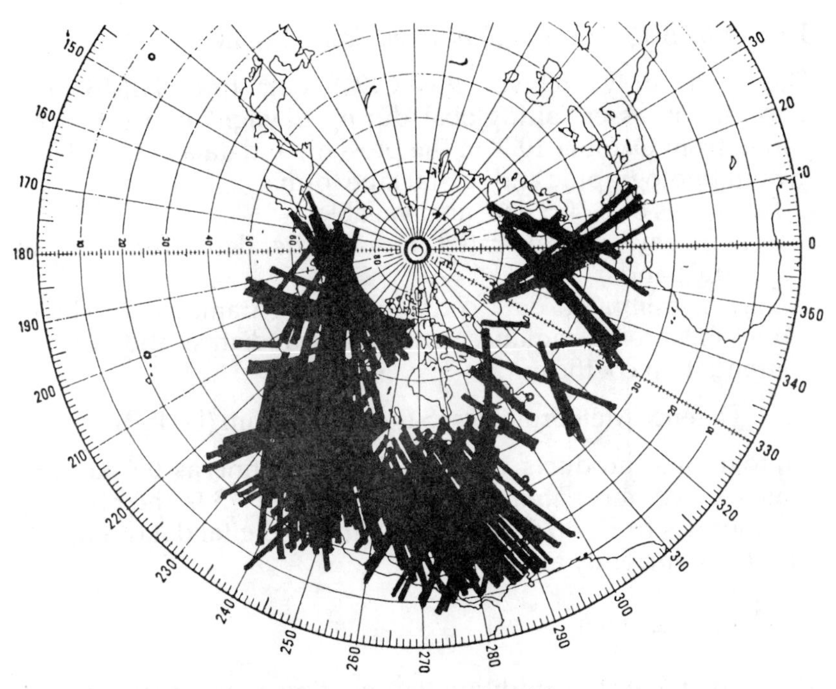

Fig. 9.4. SEASAT SAR optically correlated data coverage map.

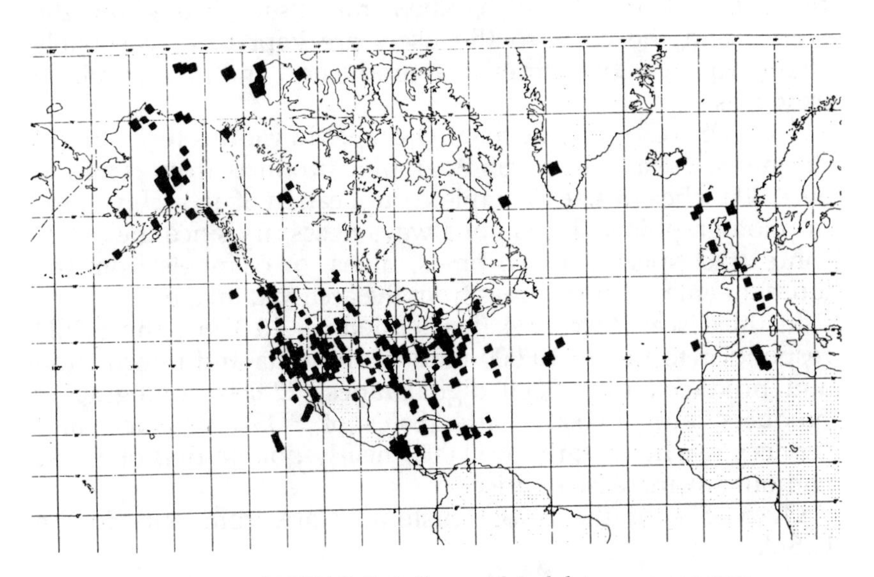

Fig. 9.5. SEASAT SAR digitally correlated data coverage map.

3.3.3. NIMBUS-7 Coastal Zone Color Scanner (CZCS)

The SDSD only archives this CZCS data from NIMBUS-7. Coverage is essentially global during daylight hours for the period from October 1978 to the present, and data are available in both photographic and digital tape format.

Two levels of CZCS data are available:

Level I: raw, uncalibrated radiometric data from all six spectral bands;

Level II: calibrated data containing the parameters of chlorophyll, aerosol radiance, subsurface radiance and diffuse attenuation coefficient.

3.3.4. Defence Meteorological Satellite Program (DMSP)

Photographic products with resolution as fine as 0.6 km are available over most areas of the globe from 1973 to the present, and mosaics from 1975 to the present. No digital data from the DMSP are currently archived.

3.4. User Services

The staff at SDSD, in addition to providing a safe archive for all these data, also responds to user requests for duplication of these data. Requests are handled from users throughout the world, and the costs to the users are kept to a reasonable charge to cover the retrieval, reproduction and dissemination expenses.

The SDSD can provide magnetic tapes in various formats and density sizes, as well as photographic products from 16mm to 30 × 30 inches in size. Complete duplication of an archive tape or photographic image is not always necessary since the SDSD can extract specific areas, times, dates, or channels from the tapes, or enlarge small areas from the archived image.

In the Fiscal Year 1984 (just ended on 10/31/84) the SDSD responded to nearly 10 000 requests for data and information and produced over 5000 digital tapes, 22 000 photographic products, and 40 000 charts. Sales of SDSD products and services reached nearly $950 000, nearly double that of FY 82 without any increase in price.

Over 25% of the SDSD customers are from outside the U.S.A.

4. APPLICATIONS OF REMOTELY SENSED DATA

While primarily intended for use by meteorologists, the NOAA environmental satellites, and the newer applications-oriented satellites and sensors, are utilized for a broad range of applications. In addition to the routine use by meteorologists and climatologists in weather mapping, forecasting etc.; these satellites and sensors are now assisting scientists, engineers, planners and others in many various fields.

During a recent survey of the SDSD users, it was discovered that nearly 25% of the data requests were of oceanographic nature and over 15% of the data requests were agriculturally oriented. One of the most recent and rapidly growing uses of these data is by law firms who utilize the data in legal cases involving airplane or ship accidents or other weather related incidents.

The wide range of users: from simple requests for a satellite image over someone's hometown, or at the time and date of a child's birth to serious multi-date or multi-year data sets for climatological studies is encountered (see Table 9.9).

In 1983 and 1984, the SDSD provided data and information for the following specific applications in addition to many others:

- U.S.A. Space Shuttle Support: 15 min data from GOES to monitor weather conditions over recovery sites;
- Airline Flight 007: to assist in search and rescue activities;

Table 9.9: Some Applications of Remotely Sensed Data.

Climatology	Geology/Minerals/Oil & Gas Faults
Meteorology	Hydrology
Severe storms	Snow Cover
Oceanography/Limnology	Glaciers
Sea Ice	Agriculture
Currents	Vulcanology
Sea Surface Temperature	Natural & Man Made Disasters
Chlorophyll	Litigation
Sediments	Search & Rescue
Waves/Tides	Education
Fisheries	Ship Routing
Geodesy	Advertising

- El Niño: satellite derived SST data to support investigation;
- Food Shortage Alerts: to monitor potential drought in 72 developing countries;
- Global Mapped Vegetation Index: to monitor crop estimates;
- Fruit Frost: to monitor potentially disasterous frost outbreaks;
- Fisheries;
  - CZCS used to chart hypoxic waters;
  - AVHRR to chart SST boundaries for fishing locations;
  - CZCS and AVHRR to define ecological boundaries for territorial disputes;
  - CZCS and AVHRR to derive 'ocean-color' charts for fisheries.
- Hurricane Warnings:
- Tornado Warnings:
- Snowmelt: monitoring heavy snow pack in U.S.
- Rainfall: estimates from severe thunderstorms;
- Ice Breakup: in Yukon River;
- Forest Fires: observation of large fires, and prediction of wind gusts;
- Volcanic Eruptions: detection and monitoring;
- Ice Freeze-Up: monitor and warning;
- Sea Fog: forecast and dissipation;
- River Effluent: detection and mapping;
- Search and Rescue: GOES and AVHRR data utilized to assist in definition of possible crash sites for downed aircraft in South America.

5. SDSD INTERNATIONAL ACTIVITIES

Over the years, the SDSD, through its representation on many international organizations, has maintained liaison with many users throughout the world, including:

- Committee On Earth Observation Satellites (CEOS);
  - working group on data.
- International Satellite Cloud Climatology Project (ISCCP);
- International Satellite Land Surface Climatology Project (ISLSCP);
- World Climate Research Project (WCRP);

- Food And Agricultural Organization's World Index Of Space Imagery (WISI);
- U.N.'s Intergovernmental Oceanographic Commission (IOC);
  - International Ocean Data Exchange [IODE];
  - Remote Sensing Task Group.

6. FUTURE SERVICES AND PLANS

NOAA is constantly striving to provide better and quicker products and services to its users. While the prime purpose of NOAA's satellite service is to provide up-to-the-minute satellite data and information to the U.S. National Weather Service, it is also looking for better ways to provide information to many other, widely varied groups including:

- the scientific research community;
- the shipping industries;
- the fisheries industries;
- general public.

6.1 High Resolution Digital Data Base

In response to user demands for the highest resolution data available, NOAA commenced in 1979 to save, for an indefinite period of time, all the raw (level IB) and processed digital data from its polar orbiting satellites, which were previously only archived for 90 days, and similarly with its full resolution GOES data commencing in 1978.

6.2 Near-Real-Time Digital Data Service

In 1980, the SDSD instituted its 'near-real-time' digital service which permits the reproduction of polar orbiter digital data within 2 days of the actual overpass.

6.3 Outside Queries To SDSD Data Bases

In 1984, SDSD opened several of its data bases for outside access to users with remote terminals to query the availability of data holdings.

6.4 Upgrading Of Mass Storage-System

In 1985, SDSD will convert from its present mass-storage system to a new one which will allow more efficient data archival production and in 1987 have in existence an on-line user service function for catalogue access, data base inquiry and order placement.

6.5 Electronic Conference System

The SDSD is currently expanding its Electronic Conference System (first established under WISI) to include all major satellite operations, data archives, data producers, and user service organizations worldwide. This expansion, under the Committee on Earth Observation Satellites, will permit these organizations a means for effective, rapid communications, referral of orders to archives, and more.

6.6 On-Line User Services

In 1986, the SDSD plans to expand its users services function to permit users to query their data bases via their own remote terminals, identify availability, place orders for data, and eventually actually receive near-real-time and retrospective data transmissions directly.

6.7 Custom Product Availability

With the advent of the full resolution, all digital archive in 1985, the SDSD will soon have the capability to provide users with custom enhanced imagery or qualitative geophysical products on a one for one basis.

6.8 Future Satellites

In addition to the ongoing commitment by NOAA to maintain two operational polar orbiting satellites and two operational geostationary satellites well into the 1990s time frame, the SDSD will, in conjunction with other U.S.A. national and international organizations, have access to and act as the archive and distribution point for the satellites and sensors listed in Table 9.10.

Table 9.10: Future NOAA or cooperative satellites.

Satellite/sensor	Organization	Time frame	Parameters
GEOSAT / ALT	U.S. (JHU/APL)	1985	wind speed, wave heights
DMSP / SSMI	U.S. (DOD)	1986	wind speed, ice observations
NROSS / ALT	U.S. (DOD)	1989	wind speed, wave heights
SSMI			wind speed, ice observations
SCAT			wind velocity
LFMR			SST
ERS-1 / AMI	ESA	1989	many (synthetic aparature radar
ALT			wind speed, wave heights
ATSR			SST
TOPEX / ALT	U.S. (NASA)	1990	wind speed, wave heights
MRAD			atmospheric corrections
SPOT-3 / OCI	France (CNES)	1990	ocean colour, chlorophyll

7. SUMMARY

As stated previously, NOAA's Satellite Data Services Division represents a unique archive of data information from many different satellites and sensors accumulated over the past 25 years.

We are committed to provide an efficient, and effective service to users of satellite and information all over the world without any restrictions. We attempt to provide users with only the data they actually require for their projects and to keep the costs of this service to a minimum. If we do not have the data the users require, we earnestly try to refer them to the correct agency in the U.S.A. or abroad that does have the data needed (see Table 9.11).

The SDSD will continue to strive to improve their products, services, and liaison with the national and international satellite remotely sensed data user community. Your suggestions, comments, complaints and praise are solicited.

Table 9.11: Points of contact for data and information.

U.S. Environmental Satellites
Satellite Data Services Division
NOAA/NESDID/NCDC
World Weather Building, Room 100
Washington, D.C. 20233, U.S.A.
(Telephone no. 301-763-8111)

NASA Experimental Satellites
National Space Science Data Center
NASA/Goddard Space Flight Center
Code 601
Greenbelt, Maryland 20771, U.S.A.
(Telephone no. 301-344-6695)

U.S. Defense Meteorological Satellite
DMSP Data Archive
National Snow & Ice Data Center
Campus Box 449
University of Colorado
Boulder, Co. 80309, U.S.A
(Telephone no. 303-492-5171)

European (Meteosat) Satellite Data
Meteorological Data Management Department
European Space Operations Center
Robert-Boschstrasse #5
6100 Darmstadt, Germany
(Telephone no. 49-6151-8861)

Indian Satellite Data
National Remote Sensing Agency (NRSA)
Data Center
4 Sardar Patel Road
Secunderabad 500-003, India

Japanese Satellites
T. Saito, (Officer-in-Charge)
Meteorological Information Center
Japan Weather Association
Kaiji Center Building
5, 4-Chome, Kojimachi Chiyoda-Ku
Tokyo 102, Japan

LANDSAT
NOAA LANDSAT SERVICES
Users Services
U.S. Geological Survey
EROS Data Center
Sioux Falls, South Dakota 57198, U.S.A.
(Telephone no. 605-594-6151)

Climatic Data
User Services/NOAA
National Climatic Data Center
Federal Building
Asheville, North Carolina 28801, U.S.A.
(Telephone no. 704-259-2850)

Oceanographic Data
NOAA/NESDIS
National Oceanographic Data Center
Page Building #1
2001 Wisconsin Avenue N.W.
Washington, D.C. 20233, U.S.A.
(Telephone no. 202-634-7500)

REFERENCES

General Subjects

Behie, G. and Cornillon, P. 1981. Remote sensing, a tool for managing the marine environment: eight case studies. *University Rhode Island Tech. Rep. No. 77.*

Cornillon, P. 1982. A guide to environmental satellite data. *University of Rhode Island Marine Techn. Rep. No. 79.*

Dismachek, D., Booth, A.L. and Leese, J.A. 1980. National environmental Satellite Service catalog of products, third edition. *NOAA Technical Memorandum NESS 109*, U.S. Dept. of Commerce, Washington, April.

Hoppe, E. H. and Needham, B.H. 1979. Environmental satellites data products and service. *Edis Magazine*, U.S. Dept. of Commerce, Washington, 10, (4).

Johnson,J. D., Parmenter, F.C. and Anderson, R. 1976. *Environment satellites systems, data interpretation & applications*. U.S. Dept. of Commerce, Washington.

Needham, B.H. (Ed.) 1979-1981. *Satellite Data User's Bull.* 1, (1), January 1979; 1, (2), August 1979; 2, (1), June 1980; 3, (1), November 1981.

Needham, B.H. 1980. Satellite data handling. *Marine Technol. Soc. J.*, 14, (6).

Needham, B.H. 1980. SEASAT & NIMBUS-7 data availability. *Marine Tech. Soc. J.* 14, (6).

Needham B.H. 1981. Oceanographic satellite remotely sensed data holdings: past, present, and future. *Pro., Oceans '81, Mar. Technol. Soc.*

Schnapf, A. 1979. *Evolution of the Operational Satellite Service: 1958-1984*. RCA Electronics, Princeton, N.J.

Winston, W. and Hoppe, E.R. 1981. *National Holdings of Environment Satellite Data of the National Oceanic and Atmospheric Administration*. U.S. Dept. of Commerce, Washington.

Geostationary Satellites

Corbell, R.P., Callahan, C.J. and Kotsch, W.J. (undated) *The GOES/SMS User's Guide*, U.S. Dept. of Commerce, Washington.

Hambrick, L.N. and Phillips, D.R. 1980. Earth locating imagery data of spin-stablized geosynchronous satellites. *NOAA Technical Memorandum 111*, U.S. Dept. of Commerce, Washington.

Hunolt, G. 1978. *VISSR Digital Archive User's Guide*. U.S. Dept. of Commerce, Washington.

NOAA Polar Orbiting Satellites

Kidwell, K.B. 1981. *NOAA Polar Orbiter User's Guide*. NOAA, SDSD, Washington.

Lauritson, L. Nelson, G.J. and Porto, F.W. 1979. Data extraction and calibration of TIROS-N/NOAA radiometers. U.S. Dept. of Commerce, *NOAA Technical Memo 107*, November.

Schneider, S. R., McGinnis D.F. and Gatlin J.A. 1981. Using NOAA/AVHRR visible and near infrared data for land remote sensing. *NOAA Tech Report No. 84*, U.S. Dept. of Commerce, Washington; April.

Schwalb, A. 1978. The TIROS-N/NOAA A-G Series. *NOAA Tech Memo #95*, U.S. Dept. of Commerce, Washington, March.

Werbowetzki, A. Atmospheric sounding user's guide. *NOAA Tech Report No. 83*, U.S. Dept. of Commerce, Washington, April.

NASA Experimental Satellites

GEOS-3 Stanley, R.J. and Dwyer, M. 1980. *NASA Wallops Flight Center GEOS-3 Altimeter Data processing Report*. NASA/Wallops Flight Center, November.

NIMBUS-7

NIMBUS-7 CZCS User's Guide and Tape Documentation.

SEASAT

Ford J.P., Blom R.G., *et al.* 1980. *SEASAT Views North America, the Caribbean, and Western Europe with Imaging Radar*. JPL Pub. pp. 80-67, November.

Barkan, C.B., Hunneycott, B. *et al. An Introduction to the Interim Digital SAR Processor and the Characteristics of the Associated SEASAT SAR Imagery*. JPL Pub.pp. 81-26, April 1.

Miscellaneous

Science, 1979. 204, (4400), pp. 1405-1424.

SEASAT SASS User's Guide.

SEASAT ALT User's Guide.

SEASAT SMMR User's Guide.

10

A Future Outlook

*William M. Callicott**

ABSTRACT

The organization which is now the National Enviromental Satellite Data and Information Service began its remote sensing efforts with the launch of TIROS-1 on April 1, 1960. This vast experience has developed into the sophisticated system we operate today and serves as a framework for future planning and development. The ultimate goal is to better understand our habitat in order to improve and preserve for the future our very existence.

1. INTRODUCTION

The environmental satellites we operate today have evolved from a long line of experimental and operational satellites beginning with TIROS-1 launched on April 1, 1960. Since then over 40 environmental satellites have been successfully placed into orbit. The Earth has been under continual surveillance since the first operational meteorological satellite, ESSA-1, which was launched in February 1966 (Epstein *et al.*, 1984).

The technical evolution of these satellites has led to the highly sophisticated systems we operate today. The first views of cloud patterns surrounding the Earth, though remarkable,

* Address of the author: Dr William M. Callicott, Office of Data Processing and Distribution, NOAA-NESDIS, Federal Building 4, Suitland, Maryland 20233, U.S.A.

were crude by today's standards. Today with the Landsat 4 and 5 satellites we can view the Earth's surface with multi-channel instruments having 30 m resolution. The data presentation has evolved from crude oblique television images to precise quantitative measurements totalling almost 300×10^9 bits of information each day from the combined land and meteorological satellites. More importantly, the U.S.A. has maintained an 'open sky' policy, making data readily available to all. Today over 1000 stations in more than 120 countries receive data from less than 11 meteorological satellites. We are now investigating methods for expanding the technology in the most economical manner.

Through the efforts of the World Meteorological Organization (WMO), nations of the world have worked closely in defining the necessary conditions for optimum use of meteorological satellite systems. These efforts include applications, research, training, the development of identical transmission characteristics and operating procedures for data collection, and standard procedures for data processing and dissemination. This work has been undertaken to mitigate the difficulties and minimize the cost for users who want to exploit space-derived data. Additionally, satellite data have been exchanged internationally along with conventional observations through the WMO's Global Telecommunications System (GTS). The regular operation of weather satellites and the global acquisition and dissemination of data are prime examples of what extensive international cooperation can do.

2. WHAT WE DO TODAY

The National Oceanic and Atmospheric Administration (NOAA) is responsible for the operation of the United States' operational environmental satellites. These include the NOAA polar orbiting TIROS-N series (NOAA Satellites), the Geostationary Operational Environmental Satellites (GOES), and the Landsat land remote sensing (Landsat) satellites. Even though each one serves a different purpose, they complement one another to some degree. The polar orbiting meteorological satellites (TIROS-N) provide a twice daily global view of the Earth through its combined orbits and present quantitized measurements of the Earth's atmosphere and surface. The

GOES satellites provide repeated views (nominally every 30 min) over the same select portion of the Earth, covering an area of about 100° of longitude over a useful latitude range of 50° north to 50° south. The GOES satellites provide either images or soundings with the Visible Infrared Spin Scan Radiometer (VISSR) and Atmospheric Sounder (VAS). Both satellite systems carry data collection systems and provide direct data broadcast services. The polar satellite carries an ozone, a search and rescue system and also a solar environmental monitoring system.

The geostationary satellites provide real time continuous observations at 30-min intervals over a viewing area encompassing 50° of latitude, longitude about a fixed subpoint position along the equator. The repeated observations, in addition to being instantly available, are combined with preceding images to provide time-scale animation which provides the added dimension of cloud and atmospheric dynamics. From this, storm development, movement and dynamics can be monitored for detection and location of developing storm systems. Cloud edge displacement is used to determine wind vector values at fixed thermal layers. In addition, with the VAS sounding capability, frequent measurements of vertical thermal and moisture layers are used to compute atmospheric soundings. Because of the real time attributes of GOES, the primary mission is to support short range forecast and warning services. Current operational applications include: computerized low-level and manually derived mid- and high-level cloud wind measurements; quantitative precipitation estimates for flash flood warnings; detection of severe convective storm development and movement; hurricane classification; detection of fog and the rate of dissipation; and freeze line monitoring.

The geostationary satellites are also used to receive and re-broadcast data collection platform data which provide environmental data from remote and often unmanned locations. There are over 4600 data collection platforms currently in the system. Three of the GOES satellites are used to provide a weather facsimile broadcast service (WEFAX) which relays image sectors collected from both the polar and GOES satellites and weather analysis charts prepared by the National Weather Service. The range of WEFAX extends from western Europe to eastern Australia. Currently over 500 charts are transmitted over the system each day. (Table 10.1 lists the present GOES, instrument complement.)

Table 10.1 Comparison of sensors and systems for present and future GOES spacecraft.

	GEOS-E/H	GEOS-Next
Imaging		
Number of Channels	2 Operational 3 Prototype Operational	5 Operational
Spectral Characteristics (μm)		
Visible	0.55 – 0.75	same
Thermal i.r.	9.7 – 12.8	10.2 – 11.2
Thermal i.r.	12.3 – 13.0	11.5 – 12.5
Thermal Window	3.8 – 4.0	same
i.r. Water Vapor	6.5 – 7.0	same
Spatial resolution (km)		
Visible	1	same
Thermal i.r.	8	4
Thermal i.r.	8	4
Thermal Window	8	4
i.r. Water Vapor	8	same
Registration		
Channel to channel	not specified	10% of the IFOV
Timeliness (min)		
Earth disk	20	same
3000 × 3000 km (min)	5 (3 channels)	5 (5 channels)
sounding		
Number of Channels	12	14 (minimum)
Spatial Resolution	14 km	8 km
Timeliness		
50 Degrees of Latitude	5 h	4 h
3000 × 30000 km	2.5 h	40 min
Registration		
Channel to channel	not specified	10% of an IFOV (2% desired)
Operation	experiment time shared with imager	separate operational instrument

Table 10.1 continued

Imaging/Sounding		
Earth Location of Sensor Data:		
Absolute	10 km	2 km
Picture to Picture	5 km	2 km (1 km goal)
Communications		
WEFAX	time shared	dedicated channel
Search and Rescue	none*	operational
	* starts on GOES G and H	
Other systems		
Data Collection System	Included	same as GOES-E/H
Solar Enviromental Monitor	Included	same as GOES-E/H
Solar X-Ray Imager	None	Optional

The primary mission of the polar orbiting NOAA satellites is to provide global coverage of: vertical profiles of atmospheric temperatures (soundings); sea-surface temperatures; and the nature and distribution of storms, clouds, ice, snow and vegetation. The polar satellite applications products are more quantitative than those of the GOES, and each satellite provides a complete global look twice each day (once over the daylight portion of the Earth and the other over the night side). In addition to the Earth environmental monitors, the NOAA satellites carry a solar environmental monitor, data collection and platform location system provided by the French ARGOS, and a Search and Rescue Mission. (Table 10.2 outlines the present NOAA polar satellite instrument complement.)

Included in each day's global data base are: composite mapped cloud image mosaics for both the visible and infrared views (three views, day and night infrared and daytime visible); approximately 16 000 atmospheric soundings; between 20 000 and 40 000 sea-surface temperature observations; vegetation index maps produced from the visible and near infrared channel data; and ice-flow boundaries and sea ice conditions from the high resolution visible and infrared channels. In addition, an ARGOS system provided by the French is used to locate up to 2000 moving platforms such as bouys and balloons. Currently 539 are in operation. The new

Table 10.2. Standard instruments, services. and direct data broadcast of the current NOAA series.

Instruments

(a) Advanced Very High Resolution (AVHRR/2) (1 km resolution)

Channels	*Wavelengths (μm)*	*Primary Uses*
1	0.58 - 0.68	daytime cloud/surface mappping
2	0.725- 1.10	water delineation, ice, and snow melt
3	3.55 - 3.93	sea surface temperature, nighttime clouds
4	10.30 -11.30	sea surface temperature, day/night clouds
5	11.50 -12.50	sea surface temperature, day/night clouds

(b) TIROS Operational Vertical Sounder (TOVS) (three-sensor system)

1. High Resolution Infrared Radiation Sounder (HIRS/2) (17.4 km resolution)

Channels	*Wavelengths (μm)*	*Primary Uses*
1-5	14.95-13.97	temperature profiles, clouds
6-7	13.64-13.35	carbon dioxide and water vapor bands
8	11.11	surface temperature, clouds
9	9.71	total O_3 concentration
10-12	8.16- 6.72	humidity profiles, detect thin cirrus
13-17	4.57- 4.24	temperature profiles
18-20	4.00- 0.69	clouds, surface temperatures under partly cloudy skies

2. Stratospheric Sounding Unit (SSU) (147.3 km resolution)

Channels	*Wavelengths (μm)*	*Primary Uses*
1-3	15	temperature profiles

3. Microwave Sounding Unit (MSU) (105 km resolution)

Channels	*Frequencies (GHz)*	*Primary Uses*
1	50.31	temperature soundings through clouds
2	53.73	
3	54.96	
4	57.95	

(c) Space Environment Monitor (SEM) (measures solar particle flux at spacecraft)

1. Total energy detector (TED): solar particle intensity from 0.3 to 20 keV
2. Medium energy proton and electron detector (MEPED): protons, electrons. and ions in 30 – 60 keV range

Services

(a) SARSAT. Satellite-aided search and rescue system: cooperative international program to locate downed aircraft and ships in distress. Participants are the United States, Canada, France, the U.S.S.R., the United Kingdom, Sweden and Norway.

(b) ARGOS. Data collection and platform location system: French system; hardware provided by France and flown by NOAA. Data relayed by NOAA to the Toulouse Space Center for ARGOS processing.

Table 10.2 continued

Direct Data Broadcasts (continuous; available to any receiving station)
(a) Automatic picture transmission (APT): Visible and infrared imagery at 4 km resolution. VHF broadcasts at 137.50 or 137.63 MHz
(b) High resolution picture transmission (HRPT): Visible and infrared data at 1 km resolution. S-band broadcasts at 1698.0 and 1707.0 MHz
(c) Direct sounder broadcast (DSB): TOVS data transmitted at 136.77 or 137.77 MHz and in the HRPT stream for use in quantitative programs

generation NOAA polar satelllites carry a Search and Rescue mission operated in cooperation with the French, Canadians and Russians. So far, the system has resulted in rescues saving 312 lives.

The NOAA satellites also carry direct broadcast systems which relay the image and sounding data to receiving stations in real time. There are presently 85 HRPT stations, of which 52 are owned and operated by foreign governments. There are more than 900 APT stations in over 122 countries, with most owned and operated by foreign entities. Similarly, there are seven direct sounder broadcast stations in five foreign countries. In all three categories, more stations are being developed in many different countries (McElroy and Schneider, 1984).

The Landsat land remote sensing satellites have been in operation since 1972 and under NOAA control since 1983. The high resolution visible and near infrared channels of the Multi-Spectral Scanner (MSS) and Thematic Mapper (TM) provide 80 and 30 m resolution images of the Earth's surface. However, because of the relatively low orbit and narrow viewing angle of the teleoptics, a repeat view of the same surface area is only available every 16 days. Also, the instrument response levels are fixed to respond to low light levels in order to detect subtle surface reflectant topography thus low sun angle illumination, thick haze and clouds obscure the imagery and detracts from the utility of the data. The principal mission of the Landsat system is to provide precise cloud-free views of the Earth's surface for renewable and non-renewable resources management. Currently, a daily average of 525 MSS scenes and 360 TM scenes are scheduled from the combined Landsat 4 and 5 satellites. About 90 scenes per day of MSS and a smaller number of TM scenes are transferred to the U.S. Archives.

Data from the Landsat are transmitted directly to ground receiving stations around the world. Nine foreign nations and the European Space Agency (ESA) currently operate 11 nationally funded Landsat ground stations (Table 10.3). In addition, stations are being built or proposed for Bangladesh, Ecuador, Pakistan, the People's Republic of China, Saudi Arabia and Upper Volta. In exchange for access to transmitted data from the Landsat system, station operators pay an annual access fee of $600 000 U.S. dollars per year and a distribution fee for each product sold.

3. THE IMMEDIATE FUTURE — REMAINDER OF THE 1980's

The environmental and land satellites operated by and for the United States will change very little for the remainder of this decade. The greatest change will occur with the land remote sensing segment. As of this writing, the U.S. Government is negotiating a contract with the Earth Observation Satellite Company (EOSAT), which is a consortium of RCA Corporation and Hughes Aircraft Company, to transfer the land remote sensing programme to the private sector. The EOSAT company will agree to launch two additional Landsat-like satellites to continue the land remote sensing programme for the U.S.A. The transition arrangement is intended to cover a ten year period.

The U.S.A. Government has passed legislation (PL98-365) which mandates that the private operator continue to observe and implement U.S. international commitments and provide foreign ground stations with unenhanced data for as long as the U.S.A. Government continues as the actual owner of the remote sensing system. The Sectretary of State has the requisite authority for ensuring that private commercial Earth remote sensing activities are conducted in strict accordance with the obligations of the United States recognized international space law.

However, because Landsat satellites will be owned by a non-government private operator, the follow-on spacecraft (Landsats 6 and 7) cannot transmit data over the S-band link which is dedicated as a U.S.A. Government authorized space link. Foreign ground station operators will have to install X-band communications in order to receive Landsat data from Landsat 6 beyond. Figure 10.1 is an outline of the proposed EOSAT system (announcement by the Earth Observation Satellite Company, 1901 N. Moore Street, Arlington, Virginia, U.S.A. 22209, 703-558-4317, Telex no. 248624).

Table 10.3. Foreign Landsat ground stations.

Country	Ground station location	Operating agency	MSS Data reception and processing	TM Data reception and processing
Argentina	Mar Chiquita est. Dec. 1980	Comision Nacional de Investigaciones Espaciales (CNIE)	yes	no
Australia	Alice Springs est. Nov. 1980	Division of National Mapping, Department of Resources and Energy (DRE)	yes	no
Brazil	Cuiaba est. May. 1983	Instituto de Pesquisas Espaciais (INPE)	yes	yes
Canada	Prince Albert est. Aug. 1972	Canada Centre for Remote Sensing (CCRS)	yes	yes
European Space Agency	Fucino, Italy Kiruna, Sweden est. Apr. 1975	European Space Agency (ESA)	yes	yes
India	Hyderabad est. Jan. 1980	National Remote Sensing Agency (NRSA)	yes	yes
Indonesia	Jakarta est. Jul. 1982	Indonesian National Institute of Aeronautics and Space (LAPAN)	no (station upgrade expected by 1985)	no
Japan	Tokyo est. Jan. 1979	National Space Development Agency (NASDA)	yes	yes
South Africa	Johannesburg est. Dec. 1980	National Institute for Telecommunications, Council for Scientific and Industrial Research (CSIR)	yes	no
Thailand	Bangkok est. Nov. 1981	National Research Council of Thailand (NRCT)	yes	nc

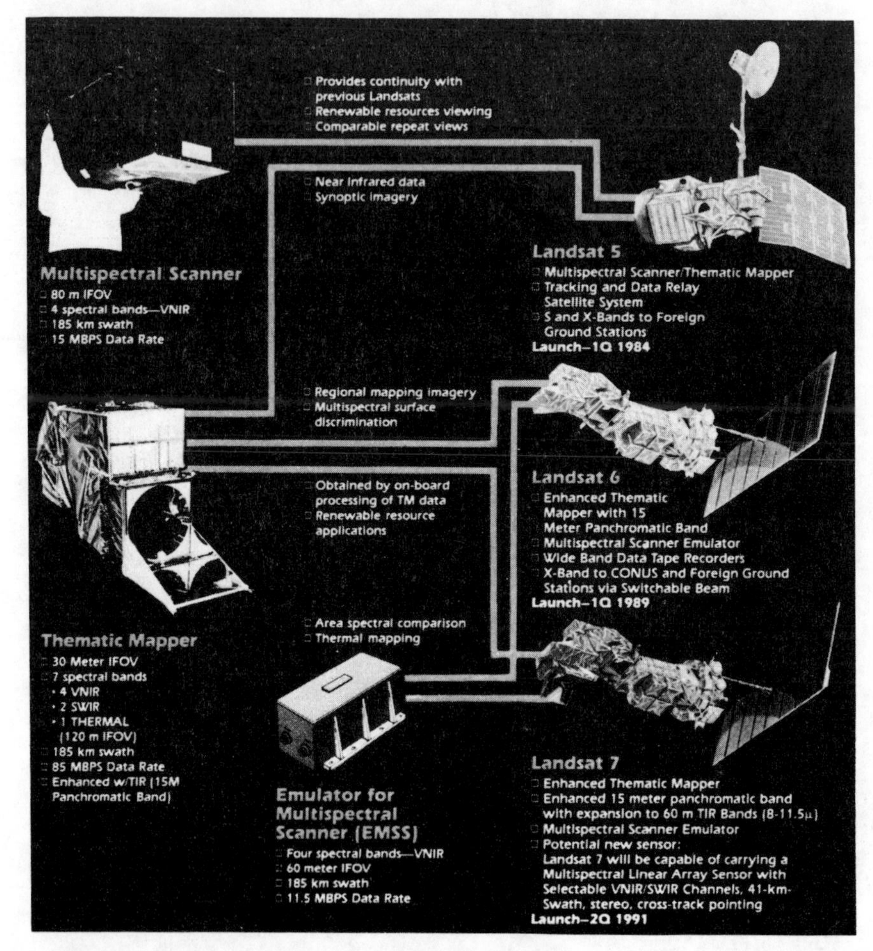

Fig.10.1 EOSAT plans

The Multispectral Scanner (MSS), flown on all previous Landsats, and the Thematic Mapper (TM), developed by the Hughes Santa Barbara Research Center (SBRC) flown on Landsats 4 and 5 have accumulated over 22 instrument-years of successful in-orbit operation and have proven the utility of remote earth observation and demonstrated the potential for commercialization.

To provide the earliest possible replacement for Landsat 5, Landsat 6 will fly an existing enhanced TM model with a 15 meter panchromatic band refurbished to flight-qualified status, and a Multispectral Scanner Emulator (EMSS) to process on-board TM data as the functional equivalent of MSS data.

Market assessments have shown a significant potential for data with higher spatial resolution and in the TIR bands. Landsat 7 will also carry on enhanced TM providing a panchromatic band with 15 m resolution and additional TIR capability in the 8-11.5 micron band.

EOSAT will continue to assess the advance high performance Multispectral Linear Array (MLA) instruments for future applications to meet the needs of the marketplace and thus provide the cornerstone of growth foreseen in the remote earth observation industry.

The current TM is an object-space line scanner that collects radiometric data in seven bands in the spectral region between 0.45 μm and 12.5 μm. The six spectral bands that sense reflected solar energy use 16 detector elements and provide 30-m ground resolution. The thermal band has four detector elements and provides a ground resolution of 120 m. There are 100 signal channels in the Thematic Mapper.

The polar satellite programme will be continued almost unchanged through the early 1990's with the addition of NOAA satellites K, L and M. The K, L and M series will bring about improvements in the sounding capabilities with the addition of a 20-channel advanced microwave sounding unit (AMSU). Fifteen of the channels are for sounding through clouds (AMSU-A) and five channels provide moisture (AMSU-B will be provided by the United Kingdom). In addition to providing soundings, the AMSU will add observations of ice beneath the clouds and observe global rain patterns. Also, the present AVHRR instrument will be altered to provide six spectral views of the Earth with a time-shared channel having 3.8 micron channel for night viewing and 1.6 micron channel for day viewing to provide distinction between snow and clouds. Table 10.4 shows the differences between the present NOAA satellites and NOAA K, L and M satellites (Miller and Sparkman, 1984).

There are many advantages to continuing the existing polar programme into the early- to mid-1990's. To begin with, the global data gathering community will benefit from a long-term collection of similar data. Also, this buys time for the United States to make a quantum jump into a future global observing system. And last, this allows the present programme to become more mature and stable with enhanced instrument performance as prior defects in design are worked out.

The future GOES programme will introduce significant operational improvements. A new VAS will be developed which will allow for concurrent imaging and dwell sounding. Also, the WEFAX will be redesigned to operate independently of the VAS for non-degraded continuous operations. The improved GOES system is called GOES-Next. The proposals call for three GOES-Next satellites with an option for two additional satellites beginning with the launch of GOES-I in 1989 carrying through to the year 2000.

New attributes of the GOES-Next aside for independent operation of the principal components of the system include an enhanced sounder having 20 versus the present 12 channels and an enhanced VISSR mode having up to 4 channels of infrared selectable. Also, the ability to Earth-locate the data to a single pixel accuracy will be a requirement.

In addition to improving the present environmental and meteorological satellite programme, NOAA is embarking on a

Table 10.4 Comparison of sensors and systems for NOAA-series, polar-orbiting satellites.

Function	*NOAA-E to -J*	*NOAA-K to -M*
Imaging	5	6
Number of channels (AVHRR)		
Spatial resolution		
Visible	0.5 km	same
i.r.	1.0 km	same
Spectral specifications		
	CH 1: 0.58 – 0.68 μm	CH 1: same
	CH 2: 0.7 – 1.1 μm	CH 2: 0.82 – 0.87 μm
	CH 3: 3.55 – 3.93 μm	CH 3a: 1.57 – 1.78 μm[a]
		CH 3b: 3.55 – 3.93 μm[a]
	CH 4: 10.3 – 11.3 μm	CH 4: same
	CH 5: 11.5 – 12.4 μm	CH 5: same
[a]Channels 3a and 3b are time shared		
Sounding		
Number of channels		
i.r. (HIRS)	23	20
Microwave (MSU, AMSU)	4	20
Spatial resolution		
i.r. (HIRS)	17.5 km (nadir)	same
Microwave (temperature)	110.0 km (nadir)	40.0 km (nadir)
Microwave (water vapor)	N/A	15.0 km (nadir)
Microwave (ice)	N/A	15.0 km (nadir)
Space environment monitor	included	no change
Ozone	SBUV (p.m. only)	SBUV (p.m. only)
Data collection system	included	increased capacity
search and rescue	included	no change
Equator crossing times		
Two satellite system	a.m. 07:30 local p.m. 14:30 local	a.m. no change p.m. 13:30 local[a]
[a]Change begins on NOAA-H		
Communications		
Automatic picture transmission (APT)	137.50 MHz 137.62 MHz	no change no change
Direct sounding broadcast	136.77 MHz 137.77 MHz	S-band at approx. 1609.5 MHz
High resolution picture transmission (HRPT)	1698.0 MHz 1702.5 MHz 1707.0 MHz	no change

comprehensive ocean observing programme. New satellites, some dedicated to ocean monitoring, will become available in 1985 with the launch of EOSAT and continue through the 1990's with subsequent launches of N-ROSS by the United States, ERS-1 by the European Space Agency, and RADARSAT by Canada. NOAA's role in ocean remote sensing is now in the planning stage with arrangements currently being negotiated for exchanging data and products through existing communication channels. The ocean programme will provide a vast increase in the quality and quantity of observations available today. The majority of user requirements will be met with surface winds, temperature, waves, sea ice, ocean-colour products and circulation provided on a global scale four times a day.

Instruments planned for ocean monitoring satellites include: a scatterometer, altimeter, SSM/I and microwave radiometer from NROSS; SAR on Radarsat; AMI, altimeter, ATSR from TOPEX; and OCI on SPOT-3. An OCI is being considered for NOAA K, L and M. Operational wind, wave and sea ice data will be increased from a present level of 2000 — 4000 reports per day to 2 000 000 plus in the 1990 time frame.

4. THE FUTURE BEYOND

Concepts are now being developed within NESDIS to support a decision to proceed with the development of a manned polar orbiting space station programme to provide a new vantage point in space from which the Earth's oceans, atmosphere and land masses can be continually monitored. There is no technical obstacle to commencing with the development of a useful polar platform that could produce dramatic advances in the practical applications of space systems by the mid 1990's. Table 10.5 outlines the characteristics possible with a two-polar platform Earth observations system.

A major attribute of a polar orbiting space station is that component failures be corrected on a component by component basis without impacting the other mission elements. Presently, when a major sensor subsystem failure occurs, a decision has to be made to call up a new spacecraft which

Table 10.5 Orbital characteristics of a two polar-platform Earth observations system.[a]

Discipline category/ instrument[b]	Afternoon platform 1.00 p.m. local time northbound[c]	Morning platform 8.30 to 10.30 a.m. southbound[c]
Atmos. and Met.	X	X
1. MRIR	X	
2. GOMR	X	
3. ERBE		
ATOVS	X	X
4. HIRS	X	X
5. AMSU		
Oceans and Ice	X	X
6. AMR	X	X
7. SSM/I	X	
8. OCI	X	X
9. ATSR	X	X
10. Altimeter	X	X
11. NSCAT	X	
12. SEASAR		
Land		X
13. MLA		X
14. GEOSAR		
Other	X	X
15. ARGOS	X	X
16. Search & rescue	X	X
17. SEM		

[a]The orbital altitude is assumed to be 800 – 1000 km.
[b]An acronym list is found before the references.
[c]Instrument totals (a.m. platform): 1765.3 watts; 795.5 kg; 273.113 mbs (includes 270 mbs for MLA and GEOSAR). (p.m. platform): 1458.7 watts; 864.6 kg; 123.914 mbs (includes 120 mbs for the SEASAR).

results in a total instrument replacement. Obviously this is not economically sound as instrument costs are only a fraction of the total spacecraft and launch costs. However, repair in orbit poses many questions which will have to be resolved, especially when determining the overall cost of the space station concept. Sufficient instrument redundancy can reduce the need for frequent costly repair missions. Fixed time preventative maintenance missions may be sufficient to maintain a continuous coverage.

Another issue to be considered is the data handling, processing and delivery. A space station with advanced instrumentation and spatial resolution will present a bulk of data far beyond today's combined remote sensing programmes. A decision will have to be made on how the data will be archived and distributed; formats and protocols will have to be established; and inter-calibration of data will have to be carefully described in order to fully utilize the potential of the data.

5. OPPORTUNITIES FOR INTERNATIONAL COOPERATION

The capabilities of the existing programmes discussed in this paper coupled with a most ambitious space station concept points out the expected growth of remote sensing in the next ten years. Remote sensing needs are beyond the capabilities and resources of any single nation. The current observations depend on systems and system components provided by a number of countries and international organizations. Such international cooperation will become even more important in the future. Because polar platforms and their payload are meant to provide a total view of the Earth's oceans, land masses, and atmosphere, international cooperation is a natural and desirable avenue for sharing benefits and costs. (Precedents for joint programmes already exist.)

As an outgrowth of the Seventh Meeting of the Economic Summit of Industrialized Nations, remote sensing was one of 18 topics chosen for discussion by expert groups. Within remote sensing, participants have discussed potential collaboration in support of polar orbiting meteorological satellites. Based on recommendations endorsed by the 1984 Summit meetings, two working groups have been formed for the coordination of activities. One of these groups, the International Polar-Orbiting Meteorological Satellite (IPOMS) group has been endorsed by the Summit to explore mechanisms for increased international cooperation in and support for polar-orbiting meteorological satellites and to ensure the continuity of these satellites. The other group is the Committee on Earth Observation Satellites (CEOS). It will provide a forum for the informal coordination of technical parameters of environmental, land and ocean satellites. CEOS membership is open to any

Table 10.6 Satellite sensors.

Satellite	Launch date	VIS/i.r. radiometer	Passive microwave radiometer	Atmospheric sounder	Radar scatter-ometer	Radar altimeter	Synthetic aperture radar	Ocean color radiometer
ITOS	1972	VHRR, SR	O	VTPR	O	O	O	O
ERTS-A	1972	MSS, RBV	O	O	O	O	O	O
DMSP	1973	SAP	O	SEE	O	O	O	O
GEOS-3	1975	O	O	O	O	RAS	O	O
NIMBUS-6	1975	O	ESMR	HIRS, THIR, SCAMS	O	O	O	O
HCMM	1978	SR	O	O	O	O	O	O
NIMBUS-7	1978	O	SMMR	THIR, SAMS	O	O	O	CZCS
SEASAT	1978	VIRR	SMMR	O	SASS	ALT	SAR	O
TIROS-N	1979	AVHRR	MSU	TOVS (HIRS/2, MSU, SSU)	O	O	O	O
LANDSAT 5	1984	MSS, TM	O	O	O	O	O	O
GEOSAT	1985	O	O	O	O	RA	O	O
SPOT	1985	HRV	O	O	O	O	O	O
IRS-1	1985	LISS	O	O	O	O	O	O
DMSP/ATN	1985	OLS	SSM/I	SSM/T	O	O	O	O
MOS-1	1986	MESSR, VTIR	MSR	O	O	O	O	O
ERS-1	1988	ATSR-M	ATSR-M	ATSR-M	AMI	E-Alt	AMI	O
N-ROSS	1989	O	SSM/I, LFMR	O	SCATT	RA	O	O
TOPEX	1989	O	O	O	O	T-Alt	O	O
NOAA-N ext	1989	AVHRR	AMSU-B	AMSU-A,B/HIRS-2	O	O	O	O
RADARSAT	1990	TBD[a]	O	O	TBD[a]	TBD[a]	SAR	O

[a]To be determined.

country with an approved remote sensing satellite programme. IPOMS membership is open to agencies contributing to the U.S.A. polar-orbiting meteorological satellites. Observer member status is extended to those countries that intend to pursue national approval for funding or in-kind contributions. The CEOS had its first meeting in Washington in September 1984, and IPOMS had its first meeting in Washington in November 1984.

6. CONCLUSIONS

The United States environmental satellite programmes will continue to operate as they are today well into the 1990's. Commitments to international cooperation will likewise continue. As the remote sensing programmes expand to provide more precise measurements over a higher density observation grid, international cooperation and sharing of costs will be necessary to spread the economies of development, acquisition, processing and distribution of data, and *in situ* observations to calibrate sensor performance.

For end users of data from meteorological, land and ocean systems there exist extensive international arrangements. Regional remote sensing facilities are located throughout Africa, Latin America and Asia assisting countries in incorporating this technology into their development plans. Organizations such as the WMO, the Food and Agriculture Organization (FAO), the U.N. Environment Program (UNEP), UNDTCD, and the United Nations Office of the Disaster Relief Coordinator (UNDRO) also make extensive use of satellite remote sensing. International scientific programmes such as NASA's Global Habitability, the International Satellite Cloud Climatology Program (ISCCP), and the International Geosphere-Biosphere Program (IGBP) will rely heavily on observations from space for worldwide monitoring. Cooperation among satellite operators will ensure that data for these activities will continue to be available in an efficient and cost-effective manner.

The impact of remote sensing from space on mankind has been tremendous. Thousands of lives have been saved through disaster (hurricanes, typhoons, drought, flooding) early warning and there has been more efficient use of financial and human capital in utilizing the Earth's resources. Cooperation in remote sensing has magnified these benefits. Such collaboration has forged important bilateral and multilateral relations (where some might not exist). Mutually beneficial ties in this area of science and technology serve national and bilateral as well as global interests.

The benefits to mankind of remote sensing from space have been demonstrated over the past 25 years. Through international cooperation these benefits will continue to be available to all nations.

ACKNOWLEDGEMENTS

The material used as a foundation for this chapter was extracted from numerous recent NESDIS papers and presentations thus, individual acknowledgements would be too extensive to list. Special thanks go to office staff, H. Brewer and S. Freeman for getting this underway, and to T. Book for preparing the layouts and J. Clapp for reviewing the section on International Cooperation.

LIST OF ACRONYMS

ALT: Radar Altimeter
AMI: Active Microwave Instrument
AMR: Advanced Microwave Radiometer (proposed by authors)
AMSU: Advanced Microwave Sounding Unit
APT: Automatic Picture Transmission
ARGOS: Data Collection System provided by French Center for National Space Studies
ATOVS: Advanced TIROS Operational Vertical Sounder (proposed by authors)
ATSR: Along-Track Scanning Radiometer — Microwave
AVHRR: Advanced Very High Resolution Radiometer
CORSS: Coordination of Ocean Remote Sensing Satellites
CLOS: Coordination of Land Observing Satellites
CZCS: Coastal Zone Color Scanner
DMSP: Defense Meteorological Satellite Program
DSB: Direct Sounder Broadcast
E-Alt: ERS-1 Altimeter
ELV: Expendable Launch Vehicle
ERBE: Earth Radiation Budget Experiment
ERBS: Earth Radiation Budget Satellite
ERS-1: ESA (European Space Agency) Remote Sensing Satellite

ERTS-A: Earth Resources Technology Satellite
ESMR: Electrically Scanning Microwave Radiometer
ESSA: Enviromental Science Services Administration
GEOS: Geodynamics Experimental Ocean Satellite
GEOSAR: Geologic Synthetic Aperture Radar (proposed by authors)
GEOSAT: Geodetic Satellite
GOMR: Global Ozone Monitoring Radiometer (proposed by authors)
GPS: Global Positioning System
HCCM: Heat Capacity Mapping Mission
HIRS: High Resolution Infrared Radiation Sounder
HRPT: High Resolution Picture Transmission
HRV: High Resolution Visible Range Instruments
JERS: Japan Earth Resources Satellite
IEOSC: International Earth Observation Satellite Committee
IGBP: International Geosphere-Biosphere Program
IPOMS: International Polar-Orbiting Meteorological Satellite Group
IRS-1: Indian Remote Sensing Satellite
ITOS: Improved TIROS Operational Satellite
LAMMR: Large Antenna Multichannel Microwave Radiometer Landsat — Land Satellite
LFMR: Low Frequency Microwave Radiometer
LISS: Linear Imaging Self Scan Cameras
MESSR: Multispectral Electronic Self Scanning Radiometer
MLA: Multispectral Linear Array
MOS-1: Marine Observation Satellite
MRIR: Medium Resolution Imaging Radiometer (proposed by authors)
MSR: Microwave Scanning Radiometer
MSS: Multispectral Scanner
MSU: Microwave Sounding Unit
NIMBUS: NIMBUS ('cloud' in Latin)
NOAA: National Oceanic and Atmospheric Administration
NOSS: National Oceanic Satellite System
N-ROSS: Navy Remote Ocean Sensing System
NSCAT: N-ROSS Scatterometer
OCI: Ocean Color Imager
OLS: Operational Linescan System
RA: Radar Altimeter
Radarsat: Radar Satellite
RAS: Radar Altimeter System
RBV: Return Beam Vidicon
SAMS: Stratospheric and Mesospheric Sounder
SAP: Sensor AVE (Aerospace Vehicle Electronics) Package
SAR: Synthetic Aperture Radar
SARSAT: Search and Rescue Satellite
SASS: Seasat-A Scatterometer Sensor
SBUV: Solar Backscatter Ultraviolet Radiometer
SCAMS: Scanning Microwave Spectrometer
SCATT: N-ROSS Scatterometer
SEASAR: Sea Synthetic Aperture Radar (proposed by authors)
Seasat: Sea Satellite

SSE: Special Sensor E
SEM: Space Environment Monitor
SIR: Shuttle Imaging Radar
SMMR: Scanning Multichannel Microwave Radiometer
SPM: Solar Proton Monitor
SPOT: Système Probatoire d'Observation de la Terre
SR: Scanning Radiometer
SSM/I: Special Sensor Micorwave Imaging
SSM/T: Special Sensor Microwave Temperature Sounder
SSU: Stratospheric Sounding Unit
STS: Space Transportation System
T-Alt: TOPEX Altimeter
TDRS: Tracking and Data Relay Satellite
THIR: Temperature-Humidity Infrared Radiometer
TIROS-N: Television and Infrared Observation Satellite
TM: Thematic Mapper
TOMS: Total Ozone Mapping Spectrometer
TOPEX: Ocean Topography Experiment
TOS: TIROS Operational Satellite
TOVS: TIROS Operational Vertical Sounder (composed of MSU, SSU, and HIRS/2)
USAF: United States Air Force
VHRR: Very High Resolution Radiometer
VIRR: Visible and Infrared Radiometer
VTIR: Visible and Thermal Infrared Radiometer
VTPR: Vertical Temperature Profile Radiometer
Windsat: Wind Satellite
WOCE: World Ocean Circulation Experiment

REFERENCES

Epstein, E.S., Callicott, W.M., Cotter, D.J. and Yates, H.W. 1984. 'NOAA Satellite Programs', IEEE Transactions on Aerospace and Electronic Systems, AES20 (4), 325-344.

Hodgkins, K.D. 1984. *International Cooperation: The Cornerstone of Remote Sensing From Space,* Pontifical Academy of Sciences — Study Week on the Impact of Space Exploration on Mankind, October 1-5, 1984, Vatican City, Rome, 28 pp.

McElroy, J.H. and Schneider, S.R. 1984. *Utilization of the Polar Platform of NASA's Space Station Program for Operational Earth Observations,* NOAA Technical Report NESDIS 12, September 1984, 67 pg.

Miller, D.B. and Sparkman, J.K. 1984. *Future U.S. Meteorological Satellite Systems,* 35th Congress of the International Astronautical Federation, October 7-13, 1984, Lausanne, Switzerland (for copy, contact the American Institute of Aeronautics and Astronautics, 1633 Broadway, New York, N.Y. 10019, U.S.A).

FURTHER REFERENCES

Baker, D.J. 1984. *Oceanography from Space, A Research Strategy for the Decade 1985-1995, Part 1* Report by the Satellite Planning Committee of the Joint Oceanographic Institutions Inc.

Barnes, J. and Smallwood, M. 1982. *TIROS-N Series Direct Readout Services Users Guide*. Enviromental Research and Technology, Inc., 120 pp.

Colwell, R.N. *et al*. 1983. *Manual for Remote Sensing*. American Society of Photogrammetry, 2 vol, 2440 p. (American Society of Photogrammetry, 210 Little Falls Street, Falls Church, Virginia, 22046 U.S.A.).

Guido, E. and Righini, M. 1980. *An S-band Receiving System for Weather Satellites,* August 1980 (contact Mr. Robert Popham, E/ER2, Satellite Programs Specialist, NOAA/NESDIS, Washington, D.C., 20233 U.S.A.).

Lauritson, L., Nelson, G. and Portor, F. 1979. Data extraction and calibration of TIROS-N/NOAA radiometers. *NOAA Tech. Memorandum NESS* 107, 58 pp.

Ohring, G., Matson, M., McGinnis, D. and Schneider, S. 1983. Applications of METSAT data in land remote sensing. *Proc. Pecora VIII Symp., Sioux Falls, S.D., October 4-7, pp. 173-186.*

Schneider, J. 1976. Guide for designing RF ground receiving stations for TIROS-N. NOAA Technical Report NESS 75, 117 pp.

Schwalb, A. 1978. The TIROS-N/NOAA A-G satellite series. *NOAA Tech. Memorandum NESS* 95, 75 pp.

Schwalb, A. 1982. Modified version of the TIROS-N/NOAA A-G satellite series (NOAA E-J). *NOAA Tech. Memorandum NESS* 116, 233 pp.

Sherman, J.W. II 1984. GEOSAT and NROSS. Proc. *18th Int. Symp. Remote Sensing Envir.*, Paris, October 1984 (in press).

11

Interpretation and Application of Spaceborne Imaging Radar Data to Geologic Problems

*Ronald G. Blom and Timothy H. Dixon**

ABSTRACT

Three synthetic aperture radars (SARs) sponsored by NASA have been flown in space. SEASAT in 1978, SIR-A in 1981 and SIR-B in 1984. Spaceborne radar image interpretation is different from airborne radar image interpretation, principally due to the wide swath and constant image geometry. Spaceborne SAR constitutes a high precision image data set in the sense that large areas are imaged under near-uniform illumination conditions. Thus spaceborne SAR images are ideally suited to geological studies of a regional nature. For spaceborne radar images a low incidence angle is probably preferable in areas of modest topography. Multifrequency observations area also needed for certain terrain-type distinctions. Subsurface imaging can occur in arid and semi-arid terrains provided that very restrictive conditions are simultaneously met. Despite the limitations sub-surface imaging has important applications in geology, hydrology, archaeology and other disciplines. Analysis of SEASAT images of a tropical terrain (Jamaica) reveals that SEASAT can provide unique information on structure and rock-type distribution on a regional scale in an environment which is not favourable for visible sensors. Possible future NASA radar missions such as SIR-B reflight, SIR-C and Space Station will provide new data from different incidence angles, frequencies and polarizations in the coming years.

* Address of the authors: Dr Ronald G. Blom and Dr Timothy H. Dixon, Jet Propulsion Laboratory, California Institute of Technology, Pasadena, Mail Stop 183-701, California 91109, U.S.A.

INTRODUCTION

This chapter reviews several aspects of the geological interpretation of spaceborne synthetic aperture radar (SAR) images. Although the interpretation methods used for airborne radar images apply, the relatively constant image geometry and synoptic coverage of spaceborne images provide new and different opportunities. We will show that some of the rules for selecting illumination geometry and other imaging parameters are not the same for spaceborne and airborne images. We have tried to collect what we feel are new potentially significant observations and results which have applications for those in the natural resource community.

Topics reviewed include the characteristics and coverage of currently available spaceborne radar image data, the effects of incidence angle and frequency, and the phenomenon of subsurface radar imaging. Examples of the utility of spaceborne radar images from two environmental extremes (tropics and desert) are given. We conclude with a review of possible future radar missions under consideration by NASA.

2. AVAILABILITY OF ORBITAL RADAR DATA

Three synthetic aperture radars have been flown in space, all built by the Jet Propulsion Laboratory (JPL) and sponsored by the National Aeronautics and Space Administration (NASA). The first, SEASAT, was a polar orbiting microwave oceanographic survey satellite launched in June 1978. The SEASAT mission was unfortunately terminated after three months due to an electrical failure in the satellite power system. The Shuttle Imaging Radar-A (SIR-A) was flown on the second flight of the Space Shuttle in November 1981. SIR-B was flown on the Space Shuttle in October 1984.

All three of these radars operated at 1.275 GHz (23.5 cm wavelengh: L-Band). The essential characteristics of these three sensors are summarized in Table 11.1, and the geographic coverage of each of the sensors in shown in Fig. 11.1. More detailed information on SEASAT ground tracks can be found in the maps in the *SEASAT Data User's Guide* (Pravdo *et al.* 1983).

SEASAT image data were restricted to regions where a ground receiving station was in view of the satellite during data

Table 11.1. The essential characteristics of the three sensors.

	SEASAT	SIR-A	SIR-B
Mission period	June – Oct, '78	Nov. 12 – 14 '81	Oct. 5 – 13 '84
Orbital inclination	108	38	57
Image coverage (sq-km).	100 million	10 million	10 million[a] 4 million[b]
Swath width (km)	100	50	15 – 60
Incidence angle (degrees)	23	47	15 – 60

[a] optical coverage
[b] digital coverage

acquisition as there was no provision for recording the SAR data onboard. The analog SEASAT downlink was digitized at the receiving station. This was subsequently optically correlated to produce long image swaths. Selected areas were also digitally correlated, which resulted in digital images representing an area of 90 × 90 km. SIR-A data were optically recorded on-board the Space Shuttle and the signal film was correlated into images on the ground after the mission. Some digital SIR-A images have been produced by scanning the film. For the SIR-B mission, both digital and optical data were obtained. The same optical film recorder used for SIR-A was used again on SIR-B. Digital SIR-B data was either transmitted directly via the Telemetry Data and Relay Satellite System (TDRSS) satellite, or recorded on digital tape on-board for later playback. Problems with the communications antenna on the Space Shuttle prevented the acquisition of much digital data. Additionally most SIR-B data acquired from larger incidence angles (greater than 45°) suffers from a low signal-to-noise ratio. The power radiated by the SIR-B antenna was lower than expected due to a transmission line problem. This problem affected both digital and optical data.

Additional information on the availability of SEASAT data is given in Holt (1981) and for SIR-A in Holmes (1983) (see also Appendix). At the time of writing SIR-B data are still being processed, but data will eventually be stored at the National Space Science Data Center (NSSDC), as for the SIR-A data.

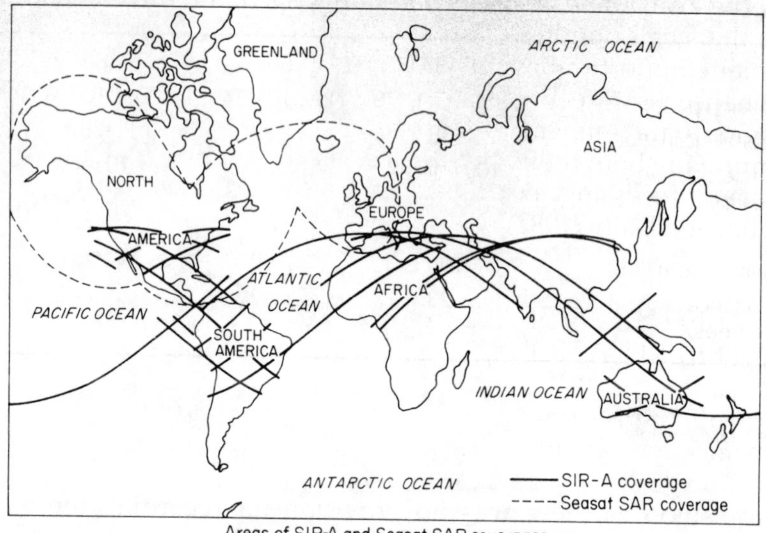

Areas of SIR-A and Seasat SAR coverages

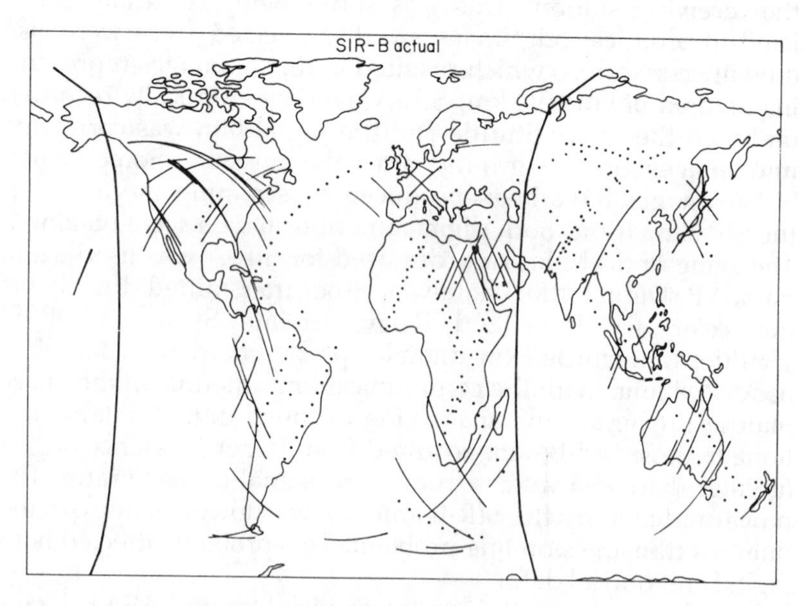

Fig. 11.1. Maps showing the SEASAT and SIR-A coverage (upper) and SIR-B coverage (lower).

The availability of digital SEASAT and SIR-B images, as well as the small number of digitized SIR-A images presents some unique opportunities for geological research. The same image enhancement techniques applied to other two-dimensional digital data sets can be applied to radar images, frequently improving both their appearance and utility. A review of digital image processing techniques useful for radar can be found in Blom and Daily (1982)

3. ASPECTS OF SPACEBORNE RADAR IMAGE INTERPRETATION

It is not our purpose here to recapitulate the essentials of radar image interpretation as there are excellent summaries available; for an excellent introduction to radar image interpretation see Sabins (1976), MacDonald (1980) and also the appropriate chapters in the new *Manual of Remote Sensing* (Colwell, 1983). Our intent here is to review and give examples of some of the less obvious aspects of orbital radar image interpretation, advantages and potential pitfalls. There are significant differences in orbital radar and airborne radar images which must be considered in any interpretation: principally image geometry and swath width.

The primary variables which control radar image tone (brightness) are, in order of decreasing effect, slope, roughness and dielectric constant (including the effect of soil moisture). Each of these variables gives information valuable to the geologist.

In the following paragraphs we shall consider the effects of incidence angle, frequency and dielectric constant of surficial materials (the latter can lead to sub-surface imaging in arid terrains). The first two items are radar system variables, and are thus potentially selectable given sufficient knowledge. The moisture level control is an environmental factor with limited but none the less very important applications.

3.1. Effects of Incidence Angle

Because orbital radar images are acquired at a great distance from the terrain (as opposed to airborne radars) the angle subtended by even a 100 km swath is quite small. In the case of

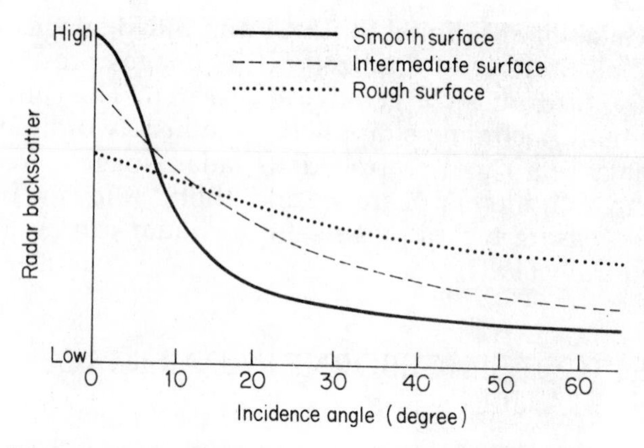

Fig. 11.2. Schematic radar backscatter curves showing relative echo strength as a function of incidence angle. Note the relative separation of the various surfaces at the various angles.

SEASAT, the incidence angle varied only 6° from the near to far range portions of the 100 km image swath. SIR-A had a similar angular variation over its 50 km image swath (the altitude was much lower). Thus, in contrast to radar images from aircraft, the imaging geometry is quite constant across any potential study area. SIR-B had the capability to vary the incidence angle, with the goal of studying the additional information provided by observing the same area at different angles. Although SIR-B data are at present being analysed, we can observe some interesting effects by comparing images from SEASAT (23°) and SIR-A (47°). Recall that the wavelengths are the same, so when viewing a temporally invariant target the incidence angle is the principal variable.

To illustrate the general behaviour of different surfaces as a function of incidence angle, consider the schematic backscatter curves in Fig. 11.2. Note the difference in relative backscatter values for rough, intermediate and smooth surfaces. It is also important to note that the greatest differences between the surfaces are at relatively small incidence angles (say, less than 30°). The ability to image the same surface at different incidence angles therefore gives information on the roughness characteristics of the surface. Also note that an aircraft image swath which subtends 20° or more of incidence angle may be difficult to interpret from the standpoint of backscatter variations because of the incidence angle effects.

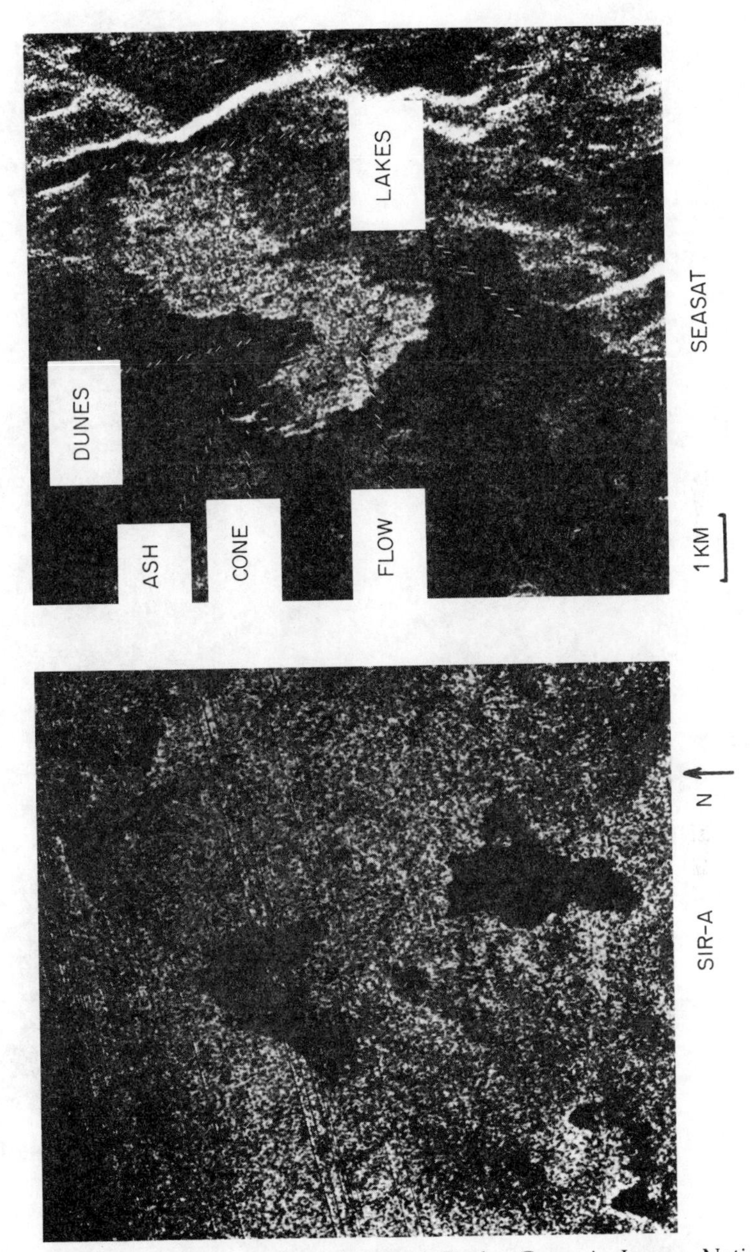

Fig. 11.3. SEASAT and SIR-A images of Cinder Cone, in Lassen National Park, CA. The principal difference between the images is incidence angle. Note that more features are discriminable at the lower incidence angle of the SEASAT image.

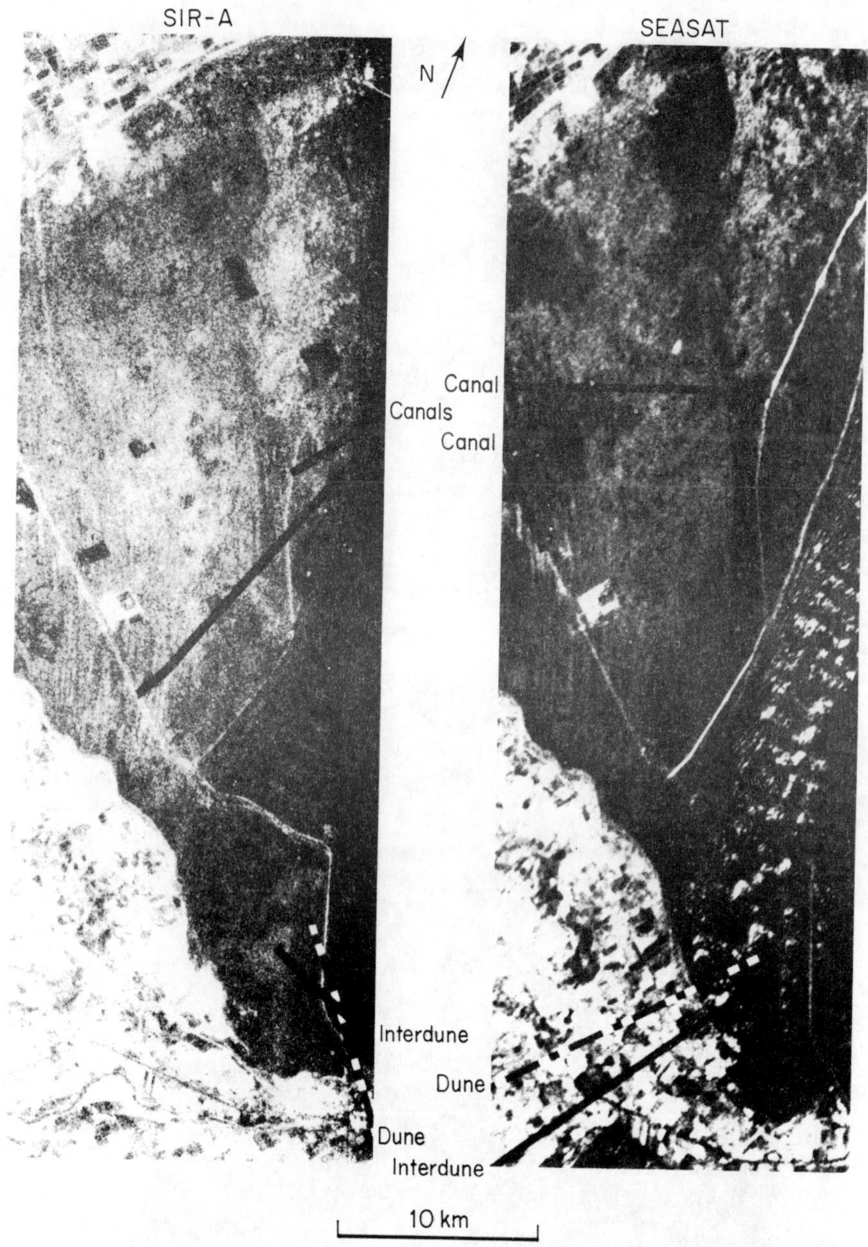

Fig. 11.4. SEASAT and SIR-A images of the Algodones Dunes. CA. In the SEASAT image, the dunes are bright while the intradune areas are darker. In the SIR-A image, the dunes are black, while the intradune flats are discernable. The dunes themselves are visible only at incidence angles below 33°.

As an example, consider the orbital radar images of a small volcanic area called Cinder Cone (part of Lassen National Park in northern California) shown in Fig. 11.3. Surfaces contained in the images are a recent lava flow (less than 200 years), a cinder cone, some ash and ash dunes, all surrounded by coniferous forest. These surface types are indicated on the SEASAT image.

In the SEASAT image it is possible to discriminate all the surfaces from one another except the ash areas and lakes. Both of these surfaces are smooth. It is possible to identify the lava flow and cinder cone by their morphology and association (note the important distinction between discrimination and identification). In the SIR-A image the forest and lava flow have become indistinguishable, as have the cinder cone and dunes *vs* the smooth ash and lakes. What was a discrimination among five differing surfaces in the SEASAT image has dwindled to discrimination between two surface types in the SIR-A image because the backscatter curves tend to merge at higher incidence angles. In a SIR-B image of this area (not shown) the same surfaces seen in the SEASAT are separable, although the dunes are less obvious. The SIR-B incidence angle in this case was about 30°. The best incidence angle for separation of the largest number of surfaces in this example is certainly less than in the SIR-A case. Previous experiment with radar images of sand dunes (Blom and Elachi, 1981) indicates that the incidence angle must be less than about 33° in order to distinguish the dunes from the other smooth surfaces.

As a second example consider the SEASAT and SIR-A images of the southern portion of the Algodones Dunes (southeastern California) in Fig. 11.4. This portion of the dune field consists of large barchanoid dunes enclosing areas of desert floor (intradune flats). In the SEASAT image the sand dunes are quite bright and easily distinguished from the darker image tone of the desert floor. In the SIR-A image, however, the dunes have become black and the desert floor is still dark, but brighter than the dunes. Thus there has been a tone reversal between these two surfaces at the two incidence angles.

The backscatter behaviour of the sand dunes is easy to understand if one considers that a dune is actually a very smooth (to the radar) surface with a non-zero slope. The bright image tone in the SEASAT image is due to specular reflection

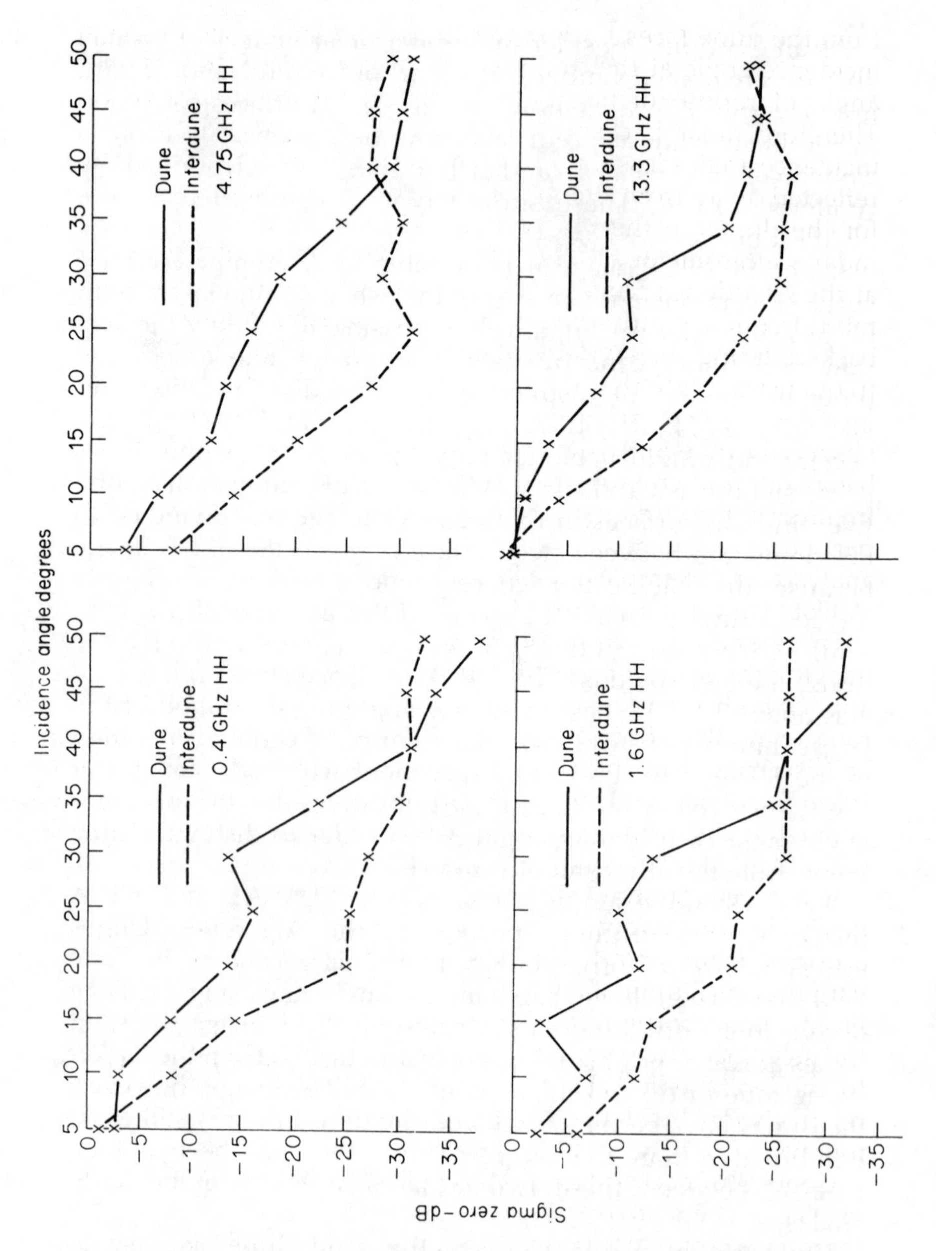

Fig. 11.5. Four frequency like polarized scatterometer curves for the dunes and intradune flats at Algodones Dunes. Note that there is a cross-over in backscatter at about 35° in all frequencies. At high angles the flats are brighter than the dunes, and vice versa at lower angles.

from the dune faces normal to the radar beam. Thus, the local incidence angle at the dune face is zero. Because sand has an angle of repose of about 33°, no slopes exceed this value. Therefore radar backscatter from a dune is possible only at incidence angles less than 33°. Otherwise the echo is specularly reflected away from the radar, and the black image tone seen for the dunes in the SIR-A image results. Figure 11.5 shows radar scatterometer profiles for a sand dune and intradune flat at the Algodones Dunes. These curves were produced from a multi-frequency airborne scatterometer system and show the backscattering behaviour at various incidence angles and frequencies. Note that the backscatter curves cross over at 35° incidence angle and consequently predict the behaviour observed in the SEASAT and SIR-A images. Note also that the behaviour and cross-over angle are essentially independent of frequency (the curves are different at the various frequencies in part because the field of view varies with frequency). Further analysis and examples of the scattering behaviour of dunes is available in Blom and Elachi (1981) and Blom and Elachi (1984).

These examples serve to illustrate that information exists in imaging the same surface with multiple incidence angles. This applies both for very smooth surfaces (sand dunes) and very rough ones (lava flows). It is noteworthy that more surfaces can be discrimated in the small incidence angle images than the larger incidence angle ones. The penalty of course, is increased geometric distortion. This distortion is tolerated in regions without great relief.

It has been asserted that higher incidence angles are more desirable for mapping on radar images. In radar images acquired from aircraft this is certainly true due to the rapid variation in illumination geometry at small incidence angles. From a spaceborne radar, however, the distortion induced by the incidence angle variation is minimal. In fact, in areas of low to moderate relief a strong argument for the benefits of a relatively low incidence angle can be made. The reasoning is as follows.

Generally, backscatter curves have the shapes displayed in Fig. 11.2. There are two important aspects of these curves: first, there is a rapid change in backscatter with incidence angle at smaller angles; and secondly, there is greater separation between the surfaces at low angles. The value of the second aspect was seen in the Cinder Cone example (Fig. 11.3). The first aspect has the important corollary that small local slopes will have a significant effect on backscatter, thus highlighting the slopes. The topography highlighted in this manner can be a strong indicator of structure and recent geomorphic processes. By this method the foreslopes are highlighted while the backslopes are merely darker, rather than hidden in shadow. Also, sub-resolution features can be detected if they produce sufficiently strong echos. For example, fault scarps approximately 1 m high are detected in SEASAT images of California if favourably oriented (Muskat *et al.*, 1981; Blom *et al.*, 1984). The SEASAT images of sand dunes are actually formed by reflections from sub-resolution dune faces. In view of the above, we would argue that low incidence angle radar images contain more information about the surface when acquired from space. Also, from a practical standpoint the power requirements are almost intractable at near grazing incidence angles from space.

3.2. Effects of Frequency

It is likely that in the near future radar images from other wavelengths will be available. The MRSE (Microwave Remote Sensing Experiment) is a European X-band (3 cm) radar that was flown on the Space Shuttle with the Spacelab mission. Unfortunately the radar failed, but it will hopefully be flown again soon. Other radar missions planned by the U.S.A. Canada, Europe and Japan may cover the 3.0 cm (X-band) and 6.0 cm (C-band) in addition to the 23 cm (L-band) portions of the radar spectrum. It is therefore worthwhile to consider what information is available in geological context from different radar wavelengths. Since the only spaceborne images available are L-band, we must examine airborne radar and also multifrequency scatterometer data.

As mentioned before, the primary factors which control radar backscatter (and hence image tone) are surface slope, surface roughness and dielectric constant (primarily moisture content). Of these factors, surface roughness response is strongly frequency-dependent, the others less so. A useful measure of surface roughness is the Rayleigh roughness criterion (Peake and Oliver, 1971). This measure has been used to explain radar backscatter of fan gravels and lava flows by Schaber *et al.* (1976) and Schaber *et al.* (1980). The measure essentially describes the backscattering behaviour of the surface as a function of frequency and roughness. The formulae are:

$$\text{Smooth} \quad h < \frac{\lambda}{8 \sin \gamma}$$

$$\text{Rough} \quad h > \frac{\lambda}{4.4 \sin \gamma}$$

h = the height of surface irregularities, or surface roughness;

λ = the radar wavelength;

γ = the grazing angle between the terrain and the incident radar wave.

Table 11.2 shows how the Rayleigh criterion varies as a function of frequency. The values are computed for a 45° incidence angle. Surfaces which are smooth will reflect the radar energy away from the sensor in a specular manner and hence be recorded as a black image tone. Surfaces which are rough are strong diffuse scatterers, backscattering strongly, and thus are very bright on radar images. Intermediate values are displayed as various grey levels in the radar image. A multispectral radar image consequently has the capability of describing the roughness of the various surfaces.

Table 11.2 Raleigh roughness criteria at various wavelenghs[a].

Band	Frequency[b]	Wavelength[c]	Smooth[c]	Rough[c]
Ka	34.8	0.86	0.05	0.3
X	10.0	3.0	0.17	1.0
C	5.0	6.0	0.34	2.0
S	3.0	10	0.57	3.2
L	1.6	23	1.41	8.0
P	0.4	75	4.24	24.1

[a] In cm RMS at 45° incidence angle
[b] In GHz
[c] in cm.

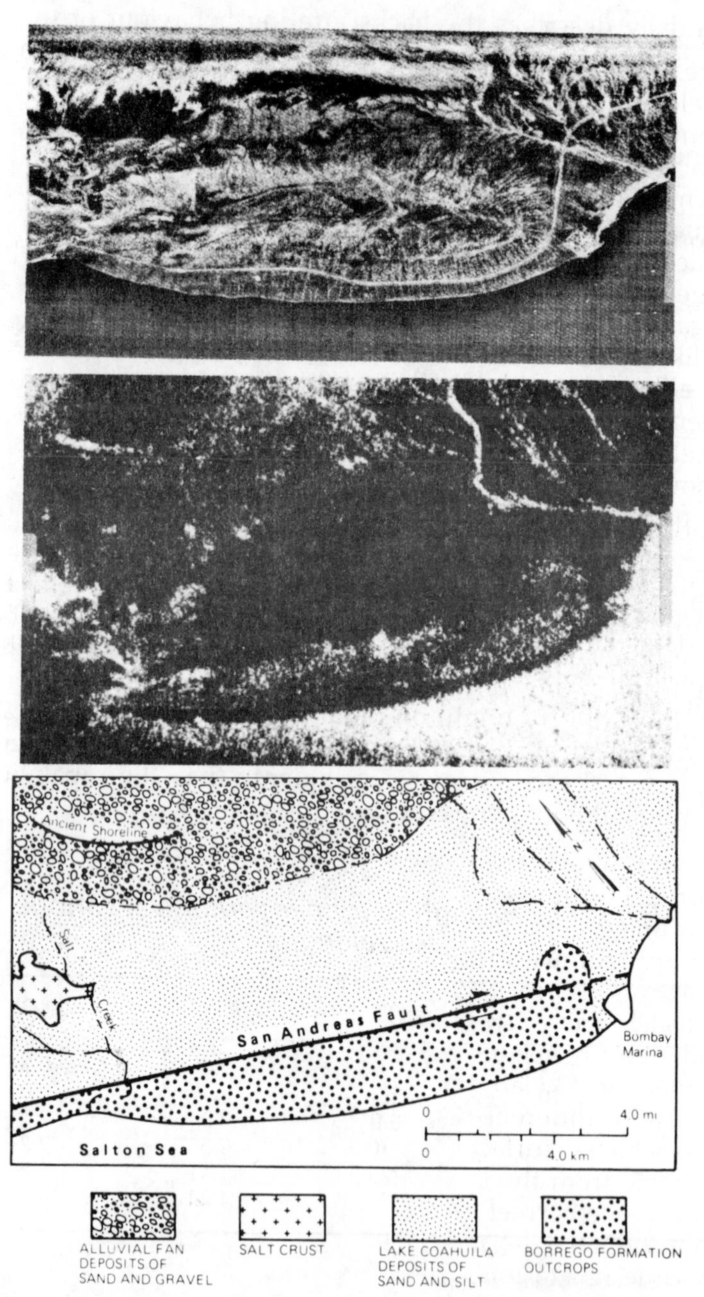

Fig. 11.6. Images of the San Andreas Fault at Durmid Hills, CA. The fault is only discernable in the SEASAT image (adapted from Sabins *et al.*, 1980).

Examining Table 11.2, it can be seen that the intermediate range covers a wider range of surface roughness at longer wavelengths. Intuitively, then, longer wavelength radars are potentially more useful for the discrimination of various surfaces because a wider range of surfaces will respond with intermediate grey levels. Experience shows that this is generally true, but other factors also influence image utility. First, any surface which is a specular reflector is recorded as a black image tone (assuming it is not inclined towards the radar as in the case of sand dunes). Lakes, dry lake beds, smooth ice and the like can therefore be confused. Context is frequently, but not always an aid. Secondly, shorter wavelength airborne systems usually have better resolution than the available spaceborne images. Thus a short wavelength radar image commonly is easier to interpret because the higher resolution and fewer smooth (specular) surfaces yields more contextural clues.

For spaceborne radars the resolution is limited by the power available to illuminate a small resolution cell and also because the data rates become prohibitively high for high resolution across a 50 or 100 km swath. (Radar resolution is a function of ability to measure time delay and doppler shift, which is independent of distance to the target. However, the measured power reflected from a resolution cell is distance–dependent and must be adequate for image formation.)

Sabins *et al.* (1980) show an instructive example of multispectral radar observations, from which the following is extracted. Figure 11.6 shows 23 cm (SEASAT) and 0.86 cm (airborne) radar images of a portion of the San Andreas Fault at Durmid Hills in Southeastern California. The San Andreas fault here juxtaposes Borrego Springs formation and Lake Cahuilla deposits. These two surface materials are compositionally similar and hence inseparable on either Landsat multi-spectral scanner or Skylab images (Sabins *et al.* 1980). However, because of differences in degree of consolidation, the roughness of these surfaces is different. The Borrego formation is separable from the Lake Cahuilla deposits in the 23 cm image because the former is moderately rough at this wavelength while the latter is smooth. Both surfaces are rough and yield the same image tone at the 0.86 cm wavelength. Despite the fact that the short wavelength image is much higher resolution, the fault is not discernable due to this lack of contrast. Thus in

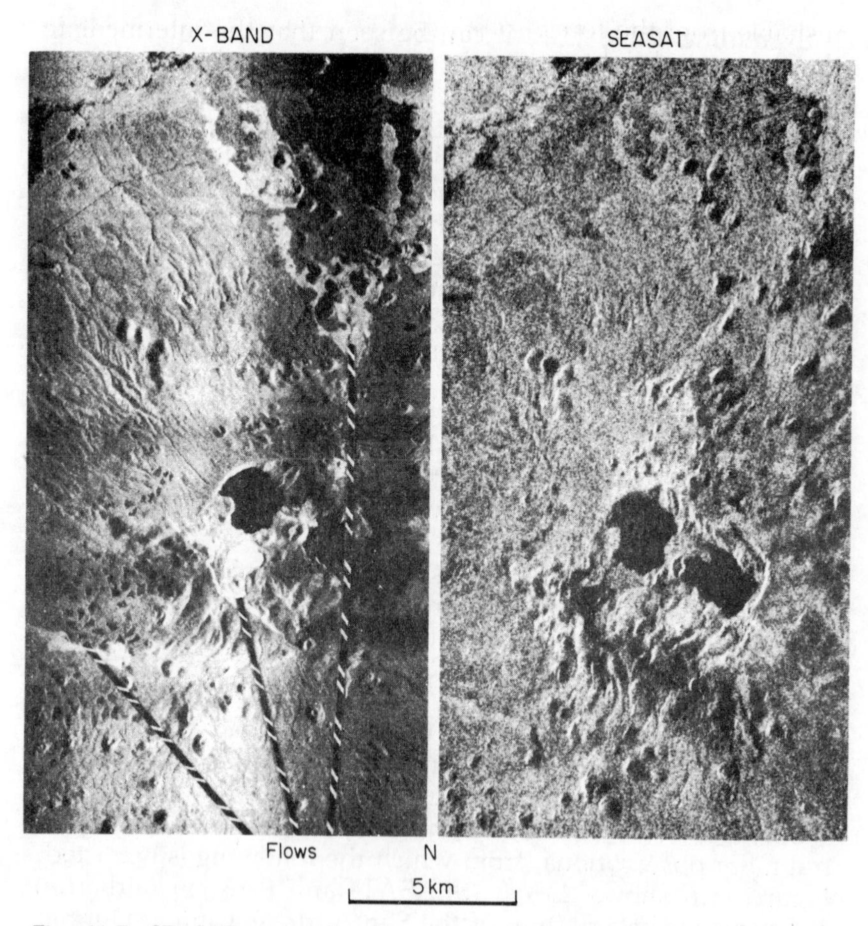

Fig. 11.7. SEASAT and X-band airborne images of Newberry Volcano, Oregon. The fresh and unvegetated lava flows have the same backscatter as the surrounding forest in the SEASAT image, yet show clearly in the X-band image.

this case the only image on which the San Andreas Fault trace can be mapped at this location in the 23 cm wavelength radar image.

Figure 11.7 shows airborne 3 cm wavelength and spaceborne 23 cm wavelength radar images of Newberry Volcano in Oregon. Recent basaltic flank eruptions and within caldera rhyolite obsidian flows are surrounded by dense coniferous forest. The lava flows are essentially completely unvegetated. In the 23 cm wavelength radar image, the flows are indistinguishable from the forest, while in the 3 cm wavelength image the flows are brighter than the forest and can be confidently mapped. To demonstrate that the inseparability at the 23 cm wavelength is not an incidence angle effect, consider the L-band scatterometer curves in Fig. 11.8. In this case it can be seen that the lava flows and the forest have remarkably similar backscatter at all incidence angles. Thus, an L-band radar image of Newberry Volcano, regardless of incidence angle, will be incapable providing a distinction between lava and forest. Figure 11.9 shows the backscatter values of these various surfaces at the four frequencies available with the scatterometer systems (at 45° incidence angle.) Note that the best separation is indicated at 13.3 GHz, very close to the X-band frequency of the airborne image in Fig. 11.8. Consequently, we here have an example where the shorter wavelength image is capable of providing information not available at the longer wavelength.

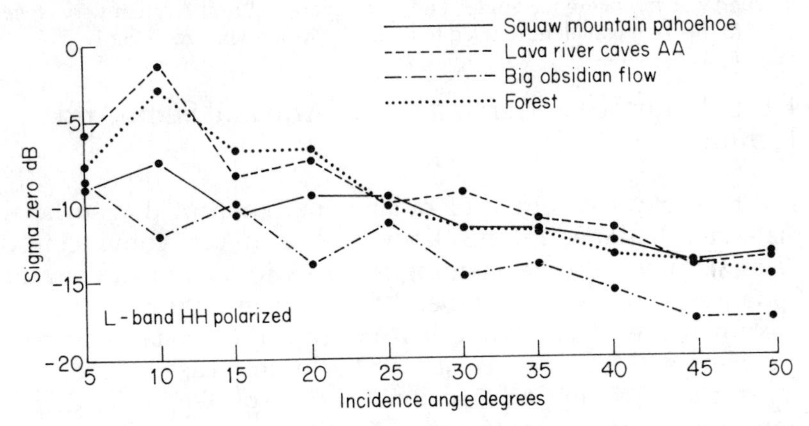

Fig. 11.8 L-band scatterometer curves for three lava flows and forest at Newberry. Note that there is little or no separation between these surfaces at this wavelength, regardless of incidence angle.

4. APPLICATIONS OF ORBITAL IMAGING RADAR DATA

Airborne imaging radar data have had many geological applications which are very well documented. The perspective from space is different in some aspects and warrants some examples. We have chosen examples from two environmental extremes to show the potential of spaceborne imaging radar data.

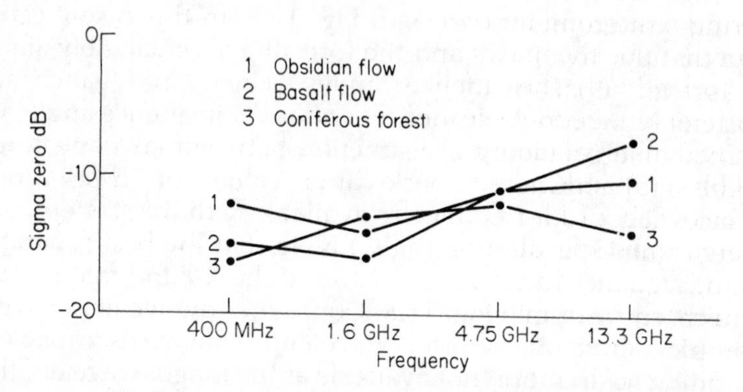

Fig. 11.9 Four frequency, like polarized, scatterometer data for surfaces at Newberry at 45° incidence angle. The best separation is at 2.25 cm very close to the 3 cm wavelength used to make the X-Band image in Fig 11.7.

4.1. Sub-Surface Radar Imaging in Arid and Semi-Arid Terrain

The theoretical potential of radar penetration of dry surficial materials has been long known, but never convincingly demonstrated. Shuttle Imaging Radar-A images of Egypt and Sudan acquired in November 1981, dramatically revealed an unknown fluvial landscape just beneath the aeolian cover of the Sahara (McCauley *et al*. 1982). In this case, there was sub-surface imaging on a regional scale. Subsequently, detection of sub-surface features in SEASAT images on a local scale in the semi-arid Mojave Desert of California was documented by Blom *et al*. (1984b). It is likely that further instances of sub-surface imaging exist in the collection of SEASAT, SIR-A and SIR-B images, and the acquisition of more orbital radar images of arid regions guarantees a wealth of sub-surface data to analyse.

The requirements for sub-surface imaging are quite strict. Review of the radar literature indicates that the following requirements must be met simultaneously in order for sub-surface imaging to occur (Blom, 1984a): (1) the surface to be penetrated must be smooth to the incident radar beam (otherwise it would backscatter and be detected by the radar); (2) the sub-surface to be detected must be rough, so that it can backscatter the incident beam; (3), the material to be penetrated must not be too thick, otherwise the signal will be attenuated during passage in and out of the material. Two to five meters appears to be the maximum depth through which an image can be formed; (4) the medium to be penetrated cannot have scatterers embedded within it. It must be homogeneous fine grained material (aeolian sand is ideal) with no cobbles or large pebbles; (5) and most importantly, the moisture content must be below about 1%. A moisture content above this causes more energy to be specularly reflected away at the upper surface/air boundary and increases the signal attenuation within the penetrated medium by raising the loss tangent of the material. The environments under which all of these criteria are simultaneously met are limited to arid regions where geologic processes produce homogeneous, fine-grained cover materials (Blom, 1984b). In retrospect then, the penetration of aeolian material in the Sahara could have been predicted.

Assuming all of the above conditions are met, there are two interesting effects which occur that increase the echo strength from the sub-surface. This first effect is that the wavelengh of the incident radar beam is actually shortened. As a consequence, a sub-surface need not be as rough as the same surface exposed to produce the same echo strength (refer to the roughness criteria in Table 11.2). The second effect is that the incoming radar beam is refracted, thus, the incidence angle is reduced at the sub-surface as compared to the surface. Referring to the backscatter curves in Fig. 11.2, one can see that an increase in backscatter occurs because of the reduction in incidence angle. Figure 11.10 shows these effects schematically. Elachi *et al.* (1984) give a complete treatment of these phenomena and calculate that under favourable conditions a sub-surface feature may in fact have a stronger echo than if it were exposed at the surface!

Figure 11.11 shows a Landsat image from the visible region and a SIR-A 23 cm wavelengh radar image of a portion of the Sahara. The Landsat image portrays a desert surface dominated by aeolian processes. Dune trains can be seen trailing across the surface. The corresponding SIR-A image shows the geomorphic expression of fluvial processes, in this case the confluence of two river systems. There is no hint of the dune

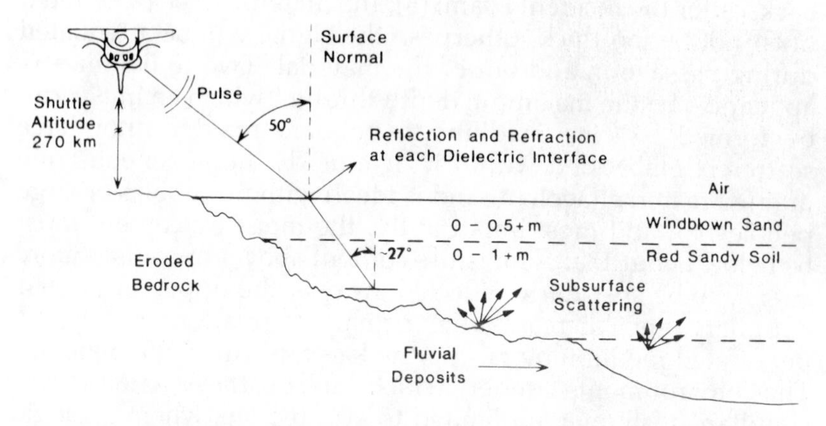

Fig. 11.10 Diagram showing refraction and reflection of the radar wave at air/surface boundary with sub-surface scattering, producing an image of the sub-surface.

trains or the sandy surface on which they move. These major tertiary rivers shrank and ultimately were abandoned as the climate deteriorated (see McCauley *et al.*, 1982, for complete details).

Recall that in the section on incidence angle effects sand dunes are imaged at incidence angle less than the angle of repose. Since the specular echo occurs even in dry sand, as in Fig. 11.10, an interesting effect can occur if an area like the Sahara is imaged at low incidence angles. Preliminary SIR-B images (30° incidence angle) of the Sahara show reflections from scattered sand dunes on the surface and where there are no dunes the sub-surface is revealed. Thus, two layers are simutaneously imaged. The difficulty, however, is that it is not always possible to tell if the radar image feature is on the surface or beneath.

One further aspect of sub-surface radar imaging must be mentioned. A common situation in many deserts is that the rock outcrops are partially submerged in windblown sand and dust. Seen from above, the view of a remote sensing device, sand may actually be the really dominant material. This can be a considerable handicap for visible and near-infrared remote sensing devices as the bulk of the signal in a given pixel will

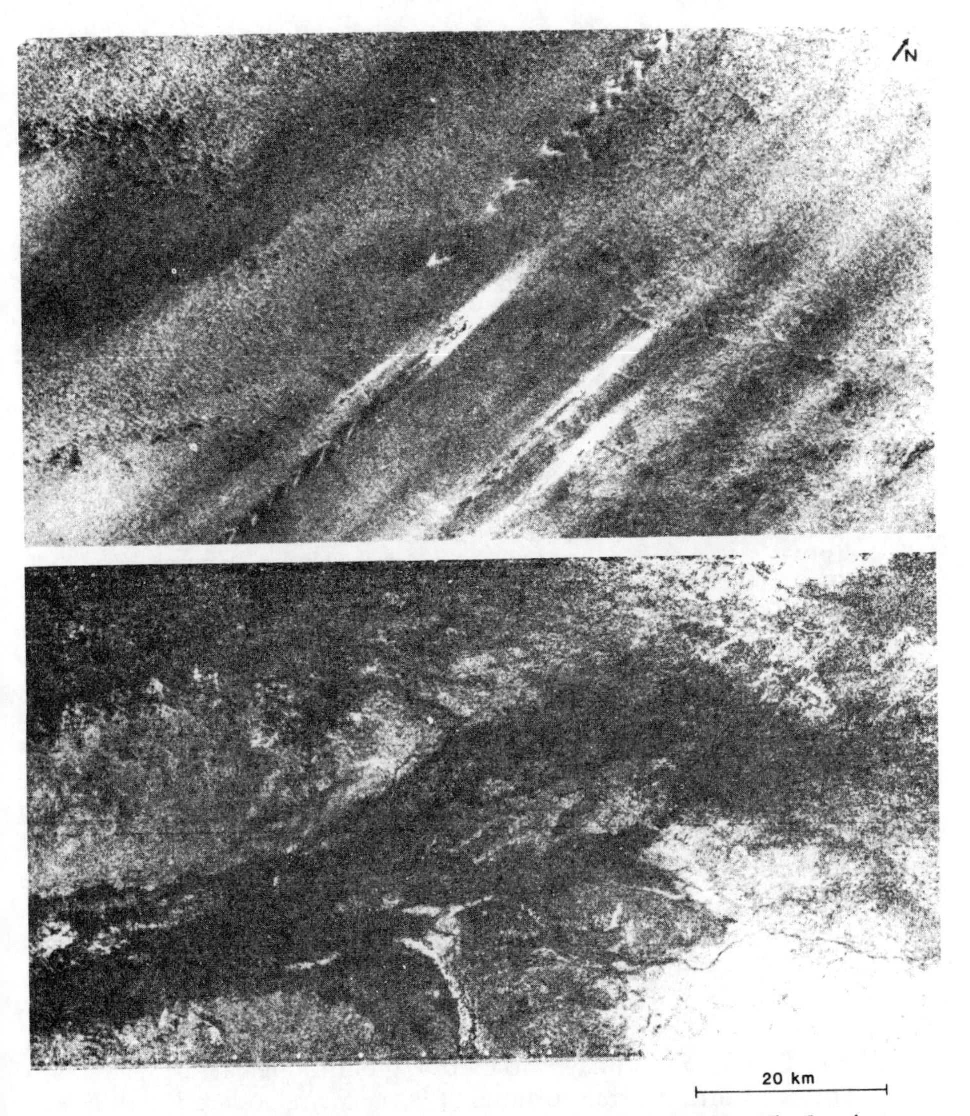

Fig. 11.11 Landsat and SIR-A images of a portion of the Sahara. The Landsat image shows an aeolian landscape while the SIR-A reveals a fluvial landscape buried just beneath. Figure adapted from Elachi *et al.* (1984).

come from the very bright sand, corrupting the signal from which one may wish to diagnose the mineralogy of the rock. In Sudan we observed that this situation was further exacerbated by the fact that the rocks are often heavily covered with desert varnish, obscuring the underlying rock. Note that the mineralogy of the varnish does not indicate the rock beneath, as the varnish derives from windblown dust and microbial activity (Potter and Rossman, 1977; Dorn and Oberlander 1981). Radar is probably the sensor of choice in this situation, because the dry sand will be transparent to the radar. In effect the radar outcrop is much better exposed than the visible outcrop. In geomorphically mature areas like northern Sudan many rock types can be discerned by the outcrop characteristics. Because radar is sensitive to roughness and highlights subtle topographic features, and because radar is not deterred by the sand cover, excellent maps can be drawn directly from the radar images once the radar signatures of the various types are known.

4.2. Geological Mapping using Seasat Images in Tropical Regions

Although tropical regions were the first obvious application of imaging radar, and significant knowledge has been gained from radar images of Brazil and Panama (among others), the synoptic view from space has provided a new perspective on the problems associated with mapping in tropical regions. For example, SEASAT gave the first unobscured large scale view of Jamaica that could be used for geological analysis, whereas all previous Landsat images were at least partially cloud covered. The high sun angle and dense vegetation cover tends to obscure diagnostic landforms. Although high resolution aerial photography and detailed geological maps were available for most of the island the SEASAT images enabled the investigation of regional structural features because of the synoptic nature of the data. (See also Sabins (1984), for an instructive application of SIR-A images to mapping in Indonesia.)

The west and central portions of Jamaica are covered with a uniformly dense vegetation cover, so that radar backscatter at L-band is largely a function of the local slope, but note that the radar probably does not penetrate the vegetation to image the surface below. At L-band the reflection is probably from volume scattering in the upper meter or two of the vegetation canopy. Because the vegetation canopy mimics the underlying topography the radar image formed from reflections of the upper part of the canopy reveal the topography below.

SEASAT SAR images proved to be an excellent tool for mapping the subtle topographic variations in the karst-dominated west and central parts of the island. Alignments of reflecting surfaces (lineaments) and textural differences among various carbonate lithologies both proved to be useful types of information in the radar images. Rather than define fundamentally new features the images allowed the recognition of new relationships among many individual features previously known from detailed group mapping.

Many of the lineaments visible in the SAR mosaic of Jamaica (Fig. 11.12) probably represent active faults because the rapid solution of the limestone under tropical weathering conditions would quickly degrade any topographic features not actively maintained. The SEASAT images allowed recognition of several new aspects of faulting in Jamaica, including a major through-going lineament (Verre-Annotto lineament; Fig. 11.13) and a series of curving scissors faults (Fig.11.12 and 11.13) in the central part of the island. Both of the features have important implications for the Neogene structural evolution of the island (see Wadge and Dixon, 1984 for complete details).

Figure 14 (a—d) shows a detailed subscene from north-central Jamaica, with lineament and texture interpretation maps as well as a geological map of the area. One of the textural units (H) is observed to undergo several kilometers of left-lateral displacement. From this and other similar observations, the left-lateral displacement on the major east–west faults on the north of the island are constrained to be less than about 10 km. Noteworthy in this figure is the much larger outcrop area of the textural unit (H) as opposed to the mapped outcrop area from the Font-Hill formation (Fig. 11.14d). This suggests that the underlying Font-Hill formation (unit Ef in Fig. 11.14d) may be controlling the erosion style of the overlying carbonate unit.

Thus, SEASAT images have contributed substantially to the geological knowledge of Jamaica. Because of the low incidence angle of the SEASAT images the topography and thus texture and structure is highlighted. The synoptic SEASAT view provided a regional picture at fairly high resolution.

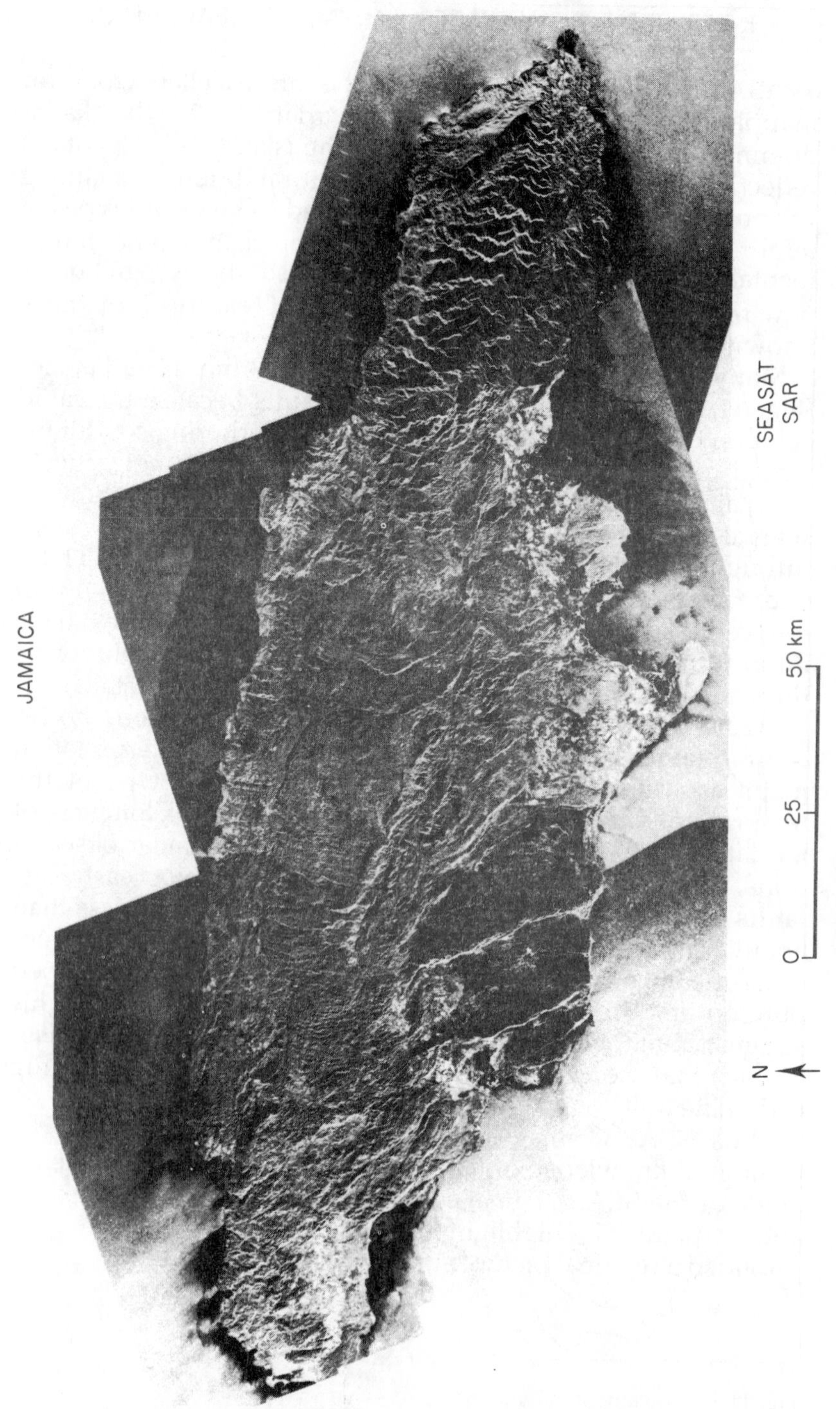

Fig. 11.12 Mosaic of SEASAT images of Jamaica. Note in particular the structural and textural content of the image. Jamaica images adapted from Wadge and Dixon (1984).

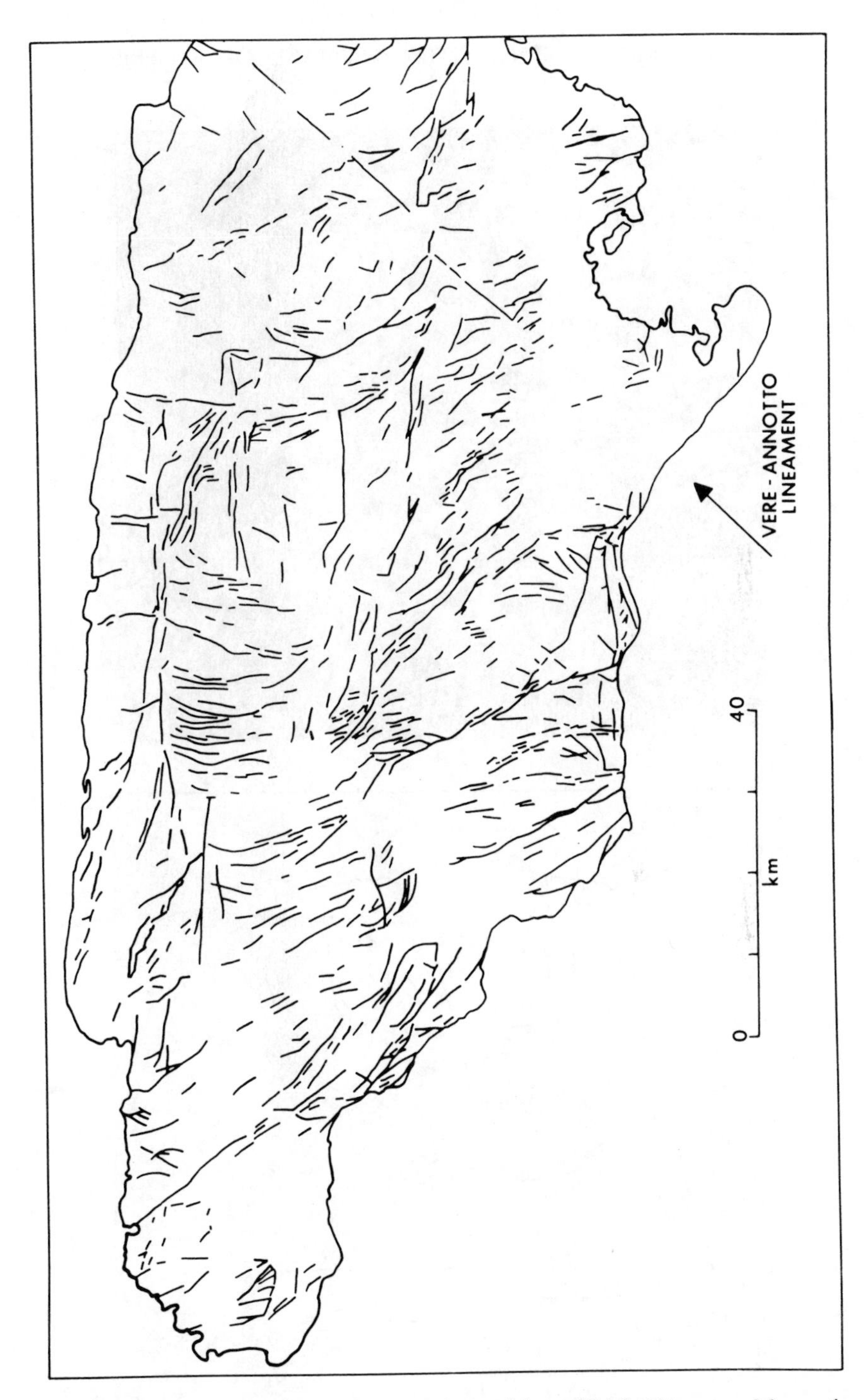

Fig. 11.13. Lineament map of Jamaica derived from SEASAT images. Many of the lineaments are known faults, others such as the Vere-Annotto lineament may be previously unkown faults and structures.

A

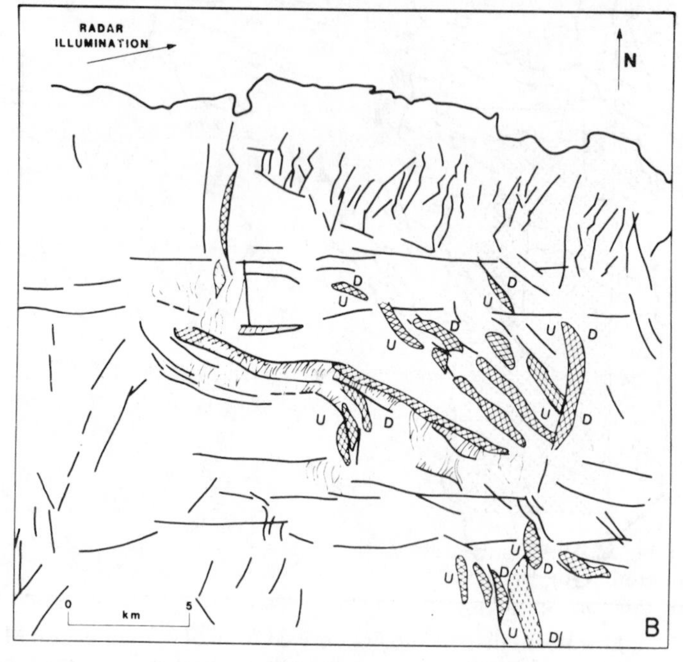
RADAR
ILLUMINATION
N
D
U
D
U
D
U
D
U
D
U
D
U
D
U
D
U
D
U
D
0
km
5
B

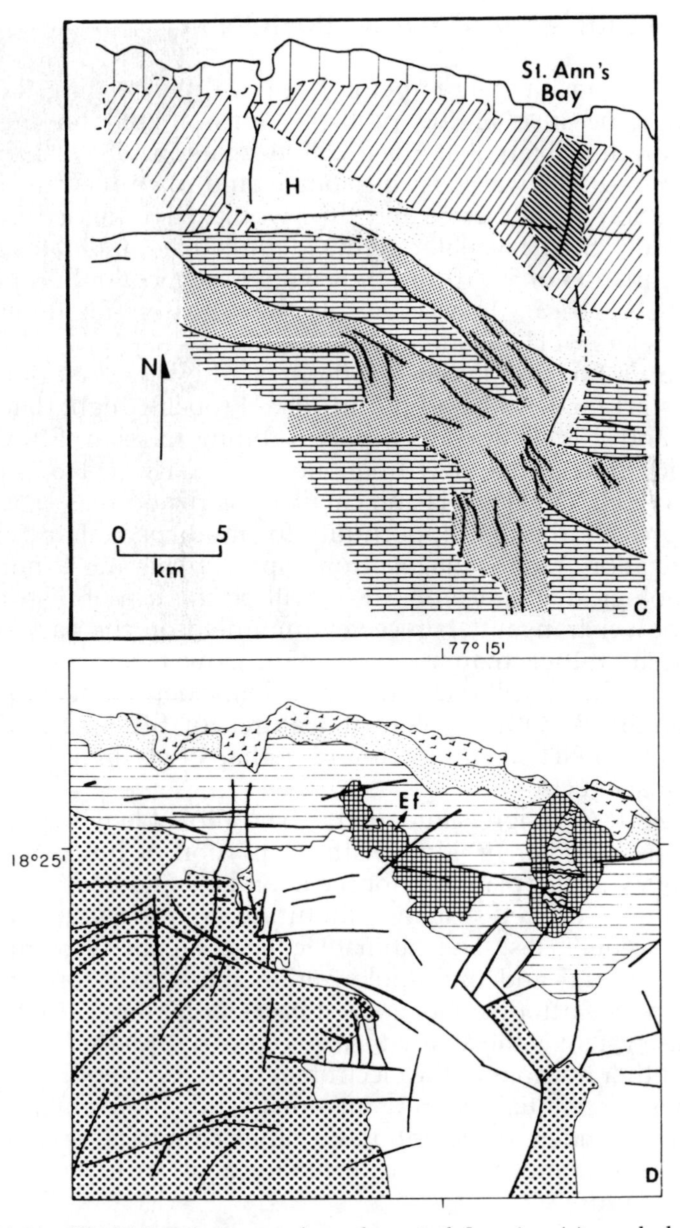

Fig. 11.14. SEASAT sub-scene of north-central Jamaica (a), and derived lineament map (b). Radar image texture units are shown in (c) and a geologic map of the corresponding area is shown in (d). Note the apparent 3 km left-lateral offset of texture unit H.

5. FUTURE NASA RADAR MISSIONS

There is considerable impetus for a Shuttle Imaging Radar-B re-flight because of the problems encountered on the first mission. It is likely that SIR-B will fly again in 1985 or 1986. An earlier re-flight was possible, but would have had an orbital inclination of only 28.5° and thus would have missed Europe and the U.S.A. Two later flights identified as candidates are at an inclination of 57 °10 and thus would see most of Europe and North America. A decision on the SIR-B re-flight and its parameters will be made soon.

Shuttle Imaging Radar-C (SIR-C) is a high priority potential new start for NASA and JPL in 1986. Probable flight date is in 1988 and it is planned as a polar orbiting mission. SIR-C will provide data at one new frequency, 4.75 GHz (C-band, 6 cm) and in HH, VV, HV, VH and circular polarizations as well. This will provide the first opportunity for multispectral and multi-polarization radar images from space. The most noteable technological advance on SIR-C will be the use of distributed low-power transmitters/receivers mounted on the back of the antenna, rather than a single high power transmitter and receiver. SIR-C will retain the variable incidence angle capability of SIR-B. Other potential features of SIR-C include an X-Band (10 GHz 3 cm) radar and perhaps two flights to study seasonal effects.

The SIR-C design is intended to be compatible with potential use on the polar orbiting platform portion of NASA's Space Station. Current plans call for the L- and C-band SAR's to fly on the initially orbited system, with future expansion capability to other frequencies. The ultimate goal is capability to provide multi-spectral and multi-polorization radar data for virtually the whole earth on an as-needed basis. A high priority feature of this system is the ability to acquire nearly simultaneous data from other portions of the electromagnetic spectrum.

These potential NASA missions and the orbital radar missions under-consideration by other countries, ensure that there will be much radar available for analysis in the not-too-distant future.

ORBITAL IMAGING RADAR SYSTEMS

	SEASAT	SIR-A	SIR-B	SPACELAB	SIR-C (2)
Launch	1978	1981	1984	1984	1985 – 6
Organization	NASA	NASA	NASA1	ESA/NASA	NASA
Band	L	L	L	X	L, C, (X)
Resolution	25 m	40 m	~20 m	~20 m	~20 M
Swath	100 km	50 m	~50 km	8.5 km	~50 KM
Look angle	20°	45°	15-60°	VARIABLE	15-60°
Polarization	HH	HH	HH	HH,VV	HH (VV)
Altitude	790°	260 km	225 km	250 km	———
Inclination	108°	38°	57°	———	POLAR ?

ORBITAL IMAGING RADAR SYSTEMS

	ERS	ERS-1	RADARSAT	SPACE-STATION
Launch	1987	1989	1990	>1990
Organization	ESA	JAPAN	CCRS	NASA
Band	C	L	C	L, C, X
Resolution	30 m	25 m	30 m	20 m$^+$
Swath	100 km	75 km	150 km	100 km
Look angle	30°	20°	30-45°	15-60°
Polarization	HH	HH	HH	QUAD
Altitude	———	———	1000 km	400 km
Inclination	POLAR	POLAR	POLAR	TBD

6. CONCLUSIONS

The radar images from the three L-Band synthetic aperture radars flown in space by NASA represent a unique and under-utilized data set. The synoptic coverage of tropics and deserts has been shown to provide new insight into the geology of both of these terrains. Sub-surface imaging provides information available in no other way for areas favourable to the technique. The constant imaging geometry and availability of multiple incidence angle data provide terrain distinctions not possible from airborne radar data. In the future multifrequency radar data will allow more surface type distinctions to be made. Spaceborne radar images should figure in any exploration effort where such data are available.

7. ACKNOWLEDGEMENTS

The engineers and scientists at JPL responsible for the development and operation of NASA's spaceborne radars are owed a considerable debt of gratitude by the user and potential user community. Radar in space has been made a reality by these individuals. The work reported here has been performed at the Jet Propulsion Laboratory of the California Institute of Technology and sponsored by the National Aeronautics and Space Adminstration.

8. APPENDIX

Inquiries about availability and purchase of SEASAT, SIR-A and SIR-B images should be directed to the following:

SEASAT: Environmental Data and Information Service, National Climatic Center, Satellite Data Services Division, World Weather Building, Room 100, Washington D.C., 20233, U.S.A.

SIR-A and SIR-B: National Space Science Data Center, Code 601, NASA Goddard Space Flight Center, Greenbelt, Maryland 20771 U.S.A.

REFERENCES

Blom, R. and Daily, M. 1982. Radar image processing for rock type discrimination: *IEEE Trans Geosci. Remote Sensing* GE-20, 343-351.

Blom, R.G. and Elachi, C. 1984 (in preparation). *Radar Scatterometry of Sand Dunes.*

Blom, R., Crippen, R. and Elachi C. 1984. Detection of subsurface features in Seasat radar images of Means Valley, Mojave Desert, California. *Geology*, 12, 346-349.

Blom, R. 1984a. *Radar Penetration Bibliography*. Unpublished JPL Report, 16 pp.

Blom, R. 1984b. Geological environments for subsurface radar imaging: a brief review. *Proc. Conf. Geolog. Problems Northeastern Africa*, South Coast Geological Society.

Colwell, R.N. (ed.), 1983. *Manual of Remote Sensing* (second Edn.). American Society of Photogrammetry: Falls Church, VA. 2417 pp.

Dorn, R.I. and Oberlander, T.M. 1981. Rock varnish: origin, characteristics and usage. *Z. Geomorph*, 25, 420-426.

Elachi C., Roth, L.E. and Schaber, G.G. 1984. Spaceborne radar subsurface imaging in hyperarid regions. *IEEE Trans. Geosci. Remote Sensing*, GE-22, 383-387.

Holmes, A.L. 1983. Shuttle Imaging Radar-A information and data availability. *Photo. Eng. Remote Sensing*, 49, 65-67.

Holt, B. 1981. Availability of Seasat synthetic aperture radar imagery. *Remote Sensing Environ.*, 11, 413-417.

MacDonald, H.C. 1980. *Techniques and Applications of Imaging* Radars, in Remote Sensing in Geology. (eds Siegal and Gillespie). John Wiley and Sons.

McCauley, J.F. *etal.* 1982. Subsurface valleys and geoarchaeology of the Eastern Sahara revealed by Shuttle radar. *Science* 218, 1004-1019.

Muskat J., Ciancanelli, E. and Blom, R. 1981. Seasat radar images for mapping in geothermal areas. *Geothermal Res. Council Trans.* 5, 115-118.

Peake, W.H. and Oliver, T. L. 1971. *The Response of Terrestrial Surfaces at Microwave Frequencies. Ohio State University Electroscience Lab, 2440-7, Tech. Rpt. AFAL-TR-70-301. Columbus, Ohio.*

Potter, R.M. and Rossman, G. R. 1977. Desert varnish: the importance of clay minerals. Science, 196, 1446-1448.

Pravdo, S.H., Huneycutt, B., Holt, M. and Held, D. N. 1983. *Seasat Synthetic Aperture Radar Data User's Manual*. JPL Publication pp. 82-90 (Write to: 'Attn. Documentation' at JPL.)

Sabins, F.F. Jr. 1978. *Remote Sensing Principles and Interpretation*. W. H. Freeman and Co. San Fransisco. 426 pp.

Sabins F.F. Jr. 1983. Geologic interpretation of Space Shuttle radar images of Indonesia. *Amer. Assoc. Petrol. Geologists Bull.*, 67. 2076-2099.

Sabins F.F. Jr, Blom, R.G. and Elachi, C. 1980. Seasat radar image of San Andreas Fault. *Amer. Assoc. Petrol. Geolog. Bull.* 64, 619-628.

Schaber, G.G. *etal.* 1976. Variations in surface roughness within Death Valley, California: geologic evaluation of 25-cm-wavelength radar images. *Geol. Soc. Amer. Bull*, 87, 29-41.

Schaber G.G., Elachi, C. and Farr, T. 1980. Remote sensing data of SP mountain and SP lava flow in north-central Arizona. *Remote Sensing Environ.*, 9, 149-170.

Wadge, G. and Dixon, T.H. 1984. A geological interpretation of Seasat-SAR imagery of Jamaica. *J. Ceol.* 92, 561-581.

List of Participants

Interregional Expert Meeting on the Use of Satellite Imaging RADAR and Thematic Mapping in Natural Resources Development, Berlin (West), 21 November–4 December 1984.

Luís Socolovsky
Chief Landsat Station
Comisión Nacional de Investigaciones Espaciales (C.N.I.E.)
Centro de Sensores Remotos
Av. Dorrego 4010
1425 Buenos Aires
Argentina

René Valenzuela R.
Director
Centre of Investigation and Application of Remote Sensing CIASER
Casilla Correo 2729
Calle Federico Zuazo 1673
La Paz
Bolivia

Hermann Kux
Associate Researcher – Microwave Remote Sensing
INPE – Instituto de Pesquisas Espaciais
Caixa Postal 515
12.200 São José dos Campos, S.P.
Brasil

R. Keith Raney
Chief Scientist
Radarsat Program,
Government of Canada
110 O'Connor Street (Suite 200)
Ottawa KIP5M9
Canada

Eduardo Díaz Araya
Director
Centro de Estudios Espaciales
Universidad de Chile
P.O.B. 5027
Santiago de Chile

He Changchui
Engineer
State Science and Technology Commission of China
The National Remote Sensing Centre (SSTCC)
San-Li-He 54
Beijing
China, People's Republic of

David Logacho
Satellite Receiving Station
Centro de Levantamientos Integrados
de Recursos Naturales por Sensores
Remotos
(CLIRSEN)
P.O. Box 8216
Quito
Ecuador

Heng L. Thung
Co-ordinator/Project Manager
Regional Remote Sensing Programme
United Nations/Econ. Social Commission
for Asia and Pacific
Rajadamnern Ave
Bangkok 10200
Thailand

H. Bodechtel
Zentralstelle für Geo-Photogrammetrie und
Fernerkundung
Luisenstr. 37
D-8000 Munich 2
Federal Republic of Germany

Hans Martin Braun
Programme Manager
Dornier System
Postfach 1360
D-7990 Friedrichshafen
Federal Republic of Germany

G. Konecny
University of Hannover
Institute for Photogrammetry
and Engineering Surveys
Nienburger Str. 1
D-3000 Hannover
Federal Republic of Germany

M. Schröder
Deutsche Forschungs- und Versuchsanstalt
für Luft- und Raumfahrt
e.V. (DFVLR)
Oberpfaffenhofen
D-8031 Wessling
Federal Republic of Germany

A. Sieber
Deutsche Forschungs- und Versuchsanstalt
für Luft- und Raumfahrt
e.V. (DFVLR)
Oberpfaffenhofen
D-8031 Wessling
Federal Republic of Germany

Gérard Brachet
Chairman and Chief Executive
Officer
Spot Image
18 Avenue Edouard Bélin
F-31055 Toulouse
France

Pierre-Henri Pisani
Centre National d'Etudes Spatiales (CNES)
Direction des Affaires Internationales et
Industrielles
2, place Maurice-Quentin
F-75039 Paris Cedex 01
France

Jean Baptiste Moussavou
Directeur de la Recherche et de la
Coopération Scientifique
Ministere de l'Enseignement Supérieur et
de la Recherche Scientifique
B.P. 2217
Libreville
Gabon

Jacub Rais
Professor and Chairman
National Co-ordination Agency for Surveys
and Mapping (BAKOSURTANAL)
P.O.B. 3546/Jkt
Jakarta
Indonesia

Victor Odenyo
Director
Regional Remote Sensing Facility
Regional Centre for Services in Surveying,
Mapping and Remote Sensing
P.O. Box 18332
Nairobi
Kenya

LIST OF PARTICIPANTS

Pil Chong Kang
Head, Remote Sensing Division
Korea Institute of Energy
and Resources (KIER)
P.O.B. Guro 98 Seoul
219-5 Garibong-Dong
Guro-Gu
Seoul 150-06
Korea

Jorge Lira
Head, Remote Sensing Department
Instituto de Geofísica
National University of
Mexico (UNAM)
Circuito Institutos - CU
04510 Mexico City - D.F.
Mexico

Jaafar Cherkaoui
Administrator
Prime Minister's Office
Palais Royal
Rabat
Morocco

Kalyan Bhattarai
Chief, Photogeology (Remote Sensing)
Section
H.M. Government of Nepal
Ministry of Industry
Department of Mines and Geology
Lainchaur, Kathmandu
Nepal

J. Neil de Villiers
Head, Future Missions and Systems Group
(Earth Observation)
European Space Agency
Keplerlaan 1
NL-2201 AZ Noordwijk
Netherlands

Simeon Omelihu Ihemadu
Director, Regional Centre for Training in
Aerial Survey
P.M.B. 5545
Ile-Ife
Nigeria

Salim Mehmud
Chairman,
Pakistan Space and Upper Atmosphere
Research Commission (SUPARCO)
P.O.B. 3125
Karachi 29
Pakistan

Ricardo M. Umali
Director General
Natural Resources Management Centre
9th floor Triumph Building
1610 Quezon Avenue
Quezon City
Philippines

Amadou Tahirou Diaw
Maître-Assistant
Université de Dakar
Département Géographie
Fac. Lettres et Sciences Humaines
Dakar
Sénégal

Lionel Seneviratne
Director, Geological Survey Dept.
Ministry of Industries and Scientific Affairs
Department of Geology
48 Sri Jinaratna Road
Colombo 2
Sri Lanka

Mohamad Chafic Safadi
Member of the National Remote
Sensing Centre Steering Committee
P.O.B. 12586
Damascus
Syria

Mustapha Masmoudi
Advisor to Prime Minister
President of Space National
National Commission
Place Kasbah Tunis
Tunis
Tunisia

Karl-Heinz Szekielda
Chief, Remote Sensing Unit
Natural Resources and Energy Division
Department of Technical Co-operation
for Development
United Nations
1 United Nations Plaza
New York, NY, 10017
USA

Jan Konecny
Prof Dr.-Ing of Photogrammetry/
Cartography
UN/DTCD
UNDP
P.O. Box 24
Mogadischu
Somalia

William Callicott
Deputy Director
Office of Data Processing and Distribution
National Environmental Satellite Data and
Information Service
Federal Building 4
Suitland
Maryland 20233
USA

Ronald Blom
Geologist
California Institute of Technology
183-701 Jet Propulsion Laboratory
Pasadena
California 91109
USA

Buzz Sellman
Manager, International Programs
Environmental Research Institute of
Michigan
P.O. Box 8618
Ann Arbor
Michigan 48107
USA

Rapporteur:
Hans-Werner Linke
Technical University Berlin
Hardenbergstr. 42
D-1000 Berlin 12
Germany

LIST OF PARTICIPANTS

German Foundation for International Development (DSE)

Volkmar Becker	Deputy Director
Herta Vomstein	Director Economic and Social Development Centre
Ulrich Böhm	Head Economic and Social Development Centre
Axel Schlarb	Programme Officer Development Planning, Surveying and Mapping Section
Elke Thurmann	Seminar Assistant
Lothar Meyer	Seminar Assistant
Christine Hoche	Programme Assistant Development Planning, Surveying and Mapping Section
Günther Mau	Seminar Assistant
Marie-Barbara von Seidlitz	Head Conference Service Section
Karin Wedeke	Deputy Head Conference Service Section
Gustavo Jaramillo	Conference Service Section
Peter Heitmann	Conference Service Section